Don Kent, PhD

EARTH DAY EVERY DAY

Dispatches From the Conservation Front

HEALTHY LANDS PUBLISHING

Library of Congress Cataloging-in-Publication Data

Kent, Don, 1955 –
Well-being and Prosperity Demand That Every Day is Earth Day: Dispatches From the Conservation Front, by Don Kent, Ph.D.
Includes bibliographical references and index.
ISBN: 978-1-7377473-5-2

1. Conservation, 2. Sustainability

DEDICATION

I thank my beloved wife, Karen Thompson, for her support and patience. Leadership is hardest with those we love the most.

TABLE OF CONTENTS

PROLOGUE

There is no us and them, here and there, just us everywhere.

Don Kent, scientist, conservationist, and leader

Who wants clean air and water? Food, medicine, and raw materials? Protection from storms, fires, and pandemics? Recreation and vacations? Earth's ecosystems services are a US$170 - $190 trillion business that doesn't bill us every month, but it does require that everyone help with maintenance. A responsibility at which we failed.

We've exceeded five of the nine planetary boundaries within which humanity can continue to develop and thrive for generations to come. Our air is unhealthy and water unpotable. Greenhouse gases are raising temperatures, changing precipitation patterns, impacting food production, and exacerbating forest fires and storms. Biodiversity, the very fabric connecting us and all other life on Earth, is diminishing.

Without exaggeration, the declining health of our planet is the greatest challenge people have ever faced. There is no immediate solution on the shelf. No technological fix waiting for someone to press the button. We've spent years fouling our air and water, cutting down forests, extracting minerals, depleting our soils, overfishing, trafficking wildlife, and destroying habitat, including our own.

Now is the time to focus attention on the health of the lands and waters that enable the ecosystem services we need for our well-being and prosperity. Now is the time to ensure that our children, and their children, have the resources needed to be well and prosperous.

How can we ensure well-being and prosperity now and into the future? It starts with education. Every one of us must understand how we are connected to the planet and everything on it – people, wildlife, plants, soil, waters, and air. We are part of one global system and dependent on all things. There is no us and them, no here and there, only us and everywhere.

We also need honest, competent, compassionate leadership, starting with ourselves. Having educated ourselves about ecosystem services and global connectedness we can take responsibility for our actions. We don't need to take the word of self-serving media or elected officials. We can turn off the media, demand honesty and competence for our elected officials, and base our decisions on the best available information.

This book is not the ultimate authority on any of the topics presented. Instead, it is an opportunity to access some of the best available information about the state of Earth's health and how it affects your life and the life of everyone else. The book also offered some clearly identified opinions to provoke your thinking, not to indoctrinate you.

You'll encounter a wide range of topics in the book. Why we need healthy lands and waters, public opinion, problems arising from climate change, poor management, and conflict over scarce resources. How to overcome the lack of trust in science, sustainable investments, and payment for ecosystem services. Environmentally smart planning, Indigenous knowledge, and holding companies accountable for destroying resources. Links between pandemics and the condition of the environment. Six articles on the critical issue of food security. And examples of how individuals and organizations are making a difference.

Above all else, thanks for reading. Enjoy the stories.

WE NEED HEALTHY LANDS AND WATERS

The health of ecosystems on which we and all other species depend is deteriorating more rapidly than ever. We are eroding the very foundations of our economies, livelihoods, food security, health, and quality of life worldwide.

Intergovernmental Science-Policy Platform on Biodiversity and Ecosystem Services 2019 Global Assessment Report on Biodiversity and Ecosystem Services.

Healthy Lands and Waters are Valuable

Without a doubt, modified lands are essential to our well-being and prosperity. Cities and towns, roadways, rail lines, dams and energy facilities, vegetable farms, and aquaculture operations meet our base needs for food, shelter, and safety, our psychological needs for belonging, love, and esteem, and our need for self-fulfillment. Modified lands are a critical element of our global economy, generating about US$80 trillion per year.

But healthy lands and waters also constitute an efficient and effective solution for meeting our base, psychological, and self-fulfillment needs. Monetarily, healthy lands are valued at US$170 - $190 trillion per year, more than twice the value of modified lands. Healthy lands provide essential goods and services and meet social, cultural, spiritual, and religious needs.

Ecosystem Services

The term used to describe the direct and indirect contributions of healthy lands and waters to human well-being is ecosystem services. Ecosystem services are conventionally categorized as regulating, cultural, habitat/supporting, and provisioning.

Regulating services moderate natural phenomena and account for 65 percent of ecosystem services value. Healthy lands and waters regulate climate, purify water, decompose waste, prevent soil erosion, moderate floods, and pollinate plants. These services are not always obvious but cannot be replaced easily or inexpensively.

Cultural services provide 18 percent of ecosystem value and consist of nonmaterial contributions to people's development and cultural advancement. For example, the role of healthy lands and waters in local, national, and global cultures, and how knowledge builds and is shared. Cultural services also include recreation and tourism.

Ten percent of ecosystem value is from habitat/supporting services. These services include photosynthesis, water and nutrient cycling, and soil creation. Services that allow the planet to sustain life and that enable regulating, cultural, and provisional services. Habitat or supporting services also include refugia and nurseries for animals and plants and maintenance of plant and animal gene pools.

Provisioning services account for 7 percent of ecosystem services value. They include consumable goods like food from land and sea, genetic material for domesticating and maintaining plants and animals, medical resources derived from plants and animals, and raw materials like forest products, natural gas, and oils.

Regardless of the ecosystem services category, we value healthy lands and waters we use now and lands we do not currently use but may want to use later. We also value lands and waters we may never use because they might benefit someone else in the future. And we value lands and waters simply because we are comforted knowing that an attractive landscape or a charismatic animal exists.

Healthy lands and waters also provide physical, social, and emotional values. They reduce our hospital stays and the number of painkilling drugs we need. They help us recover from mental fatigue and psychophysiological stress. They make us feel better by slowing our heartbeat, lowering our blood pressure, and promoting relaxed brain wave patterns. Healthy lands and waters increase self-esteem, happiness, educational status, and emotional development in at-risk children and others.

Healthy lands and waters even make our homes more comfortable. The rustle of leaves and the wind blowing through branches create a pleasing background sound and reduce noise. Homes near healthy lands and waters are cooler in the daytime and warmer in the nighttime. And when it comes time to sell, your property values are 5 to 20 percent higher than homes in less ideal settings.

The Decline of Healthy Lands

The values of both developed and healthy lands argue for a circumspect, balanced plan for global and local development and conservation of natural resources. Yet, despite our dependence on healthy lands, their destruction, degradation, and unsustainable use continue at an unprecedented rate.

Humans have changed ecosystems more rapidly and extensively over the last 50 years than in any comparable period in our existence, mostly in response to our need for food, fresh water, timber, fiber, and fuel. Ten to 20 percent of the planet's land is degraded. Fourteen of 24 ecosystem services examined by the Millennium Assessment, a comprehensive examination of ecosystems and human well-being, are degraded or degrading.

One million animal and plant species are threatened with extinction, including those domesticated for food and agriculture. Ninety percent of the world's seabirds have fragments of plastic in their stomach.

The declining health of our lands and waters impacts nearly every dimension of human health and well-being.

- Each year, one in every four deaths, more than twelve million people, are due to air, water, and soil pollution, chemical exposure, climate change, and ultraviolet light.
- Of 7.8 billion people in the world, 750 million lack adequate access to clean drinking water, and 805 million don't have enough food to eat.
- Around nine million people die of hunger and hunger-related diseases every year, many due to climate change-induced drought.
- In the Mekong River watershed alone, dams are affecting the health and livelihood of sixty million people.
- Excessive nutrient loading diminishes water quality, leading to drinking water shortages, coastal dead zones, and collapsing fisheries.
- Climate change and habitat loss are reducing biological diversity at a rate 100 to 1,000 times higher than the naturally

occurring background extinction rate – a rate expected to accelerate in the coming years.
- We are losing ecosystem services value at the rate of US$4.3–$20.2 trillion per year.

Forward-looking governments, nonprofits, businesses, and individuals understand that US$170 - $190 trillion in ecosystem services is essential for human well-being and prosperity. We don't have to give up our cities and towns, roadways, rail lines, dams and energy facilities, vegetable farms, and aquaculture operations. But we do have to sustain and make resilient our healthy lands and waters.

WHAT DOES THE PUBLIC THINK ABOUT CLIMATE CHANGE?

To all of you who choose to look the other way every day because you seem more frightened of the changes that can prevent catastrophic climate change than the catastrophic climate change itself. Your silence is worst of all.

Greta Thunberg, activist

Climate Change is Real

A 2020 Pew Research Center study examined public perception in twenty countries including Russia, Europe, the Americas, and the Asia-Pacific region. Most people surveyed in all countries believe climate change is affecting where they live (median 70 percent, range 55-86 percent), and that climate change is a serious problem (median 57 percent, range 49-80 percent). In 12 of 15 countries previously surveyed, belief in the seriousness of climate change increased by 9-25 percent.

The belief that human activity contributes a great deal to climate change is now the majority opinion in fifteen of the twenty countries examined (median 54 percent, range 34-78 percent). A majority also believe that their government is doing too little in sixteen of the twenty countries (median 58 percent, range 37-81 percent).

In the United States, most believe that climate change is affecting where they live (59 percent), is a serious problem (53 percent), and that the national government is doing too little (63 percent). Although half (51 percent) do not believe human activity contributes a great deal to climate change.

Political Ideology

Globally, differences in responses are related to political ideology. In all twenty countries, the left is more likely to believe climate change affects where they live than the right (8-51 percent difference), and that climate change is a serious problem (1-64 percent difference). Except for South Korea and the Czech Republic, the left is more inclined to believe that their national government is doing too little (2-57 percent).

Ideology is especially apparent in the United States. A large majority of the left believe climate change is affecting where they live (83 percent), the government is doing too little (92 percent), and climate change is a serious problem (86 percent). By contrast, a minority of the right believe climate change is affecting where they live (32 percent), the government is doing too little (35 percent), and climate change is a serious problem (22 percent). Despite their differences, Democrats and Republicans agree on planting trees to offset carbon dioxide emissions and a tax credit for technologies that capture and store carbon.

Worrying in Silence

Studies in 2019 by the Yale Program on Climate Change Communication and the George Mason University Center for Climate Change unearthed additional insights about what people are thinking in the United States.

- Only one in seventeen people understand the strength of the consensus among climate change scientists.
- Thirty-eight percent believe the United States is now being harmed by global warming.
- Most Americans are worried about harm from extreme heat (69 percent), droughts (64 percent), flooding (60 percent), and water shortages (59 percent).
- Six in 10 Americans rarely or never discuss global warming with family and friends.
- Five in ten strongly or somewhat support climate activists.
- Only one in three think elected officials are influenced by advocacy actions.
- Five in ten would sign a petition about global warming.
- One in three Americans have rewarded (e.g., buying products) or punished (e.g., not buying products) companies taking steps to reduce global warming.

Sacrifice, Not So Much

More than 60 percent of United State residents polled expect that climate change only requires minor sacrifices. Most adults support raising taxes on wealthy households and companies that burn fossil fuels, and a fuel-efficiency standard. They are opposed to a $2 or $10 monthly tax on their electric bill, adding to the national debt, and increasing the federal gas tax by $0.10 or $0.25 per gallon. Most (70 percent) expect that technological advances will avoid the negative effects of climate change.

Confusion About Cause

The 2019 Washington Post and Kaiser Family Foundation poll found people unclear about what causes climate change. Forty-three percent of adults and 57 percent of teenagers think plastic bottles and bags are a major contributor to climate change. Fifty-eight percent of adults cited the sun getting hotter as a major or minor contributor. Sixty percent said volcanic eruptions are a major cause of climate change.

Many are unaware that the top sources of climate change are electricity generation, transportation, agriculture, industrial production, and deforestation. Science communication experts say it is not essential for people to know the details if they understand the seriousness of the issue, the human role in solving the problem, and the need to act.

Not a Significant Factor in Voting

Despite growing concern about climate change, a September Quinnipiac poll found that voters will choose a candidate based on the economy, and to a lesser extent law and order, coronavirus, and racial inequality. Democrats, white women with four-year college degrees, and 18-34-year-old voters showed the most interest in climate change, but only 9-11 percent.

More Belief and Ideology

Another 2019 poll by the Washington Post and the Kaiser Family Foundation found a significantly higher number of United States residents (76 percent) than the Pew study (53 percent) believe climate change is a major problem or crisis. As with the Pew study, Democrats (90 percent) and independents (82 percent) are more likely than Republicans (60 percent) to say humans are causing climate change.

WHY SHOULD THE PUBLIC BE THINKING ABOUT CLIMATE CHANGE?

Saving our planet, lifting people out of poverty, advancing economic growth... these are one and the same fight. We must connect the dots between climate change, water scarcity, energy shortages, global health, food security, and women's empowerment. Solutions to one problem must be solutions for all.

Ban Ki-moon, former Secretary General of the United Nations

The public is affected by climate change regardless of their beliefs or the cause. The Intergovernmental Panel on Climate Change projected that drought, extreme temperature, flood, storm, and wildfires would increase beyond a 1.5℃ rise in temperature. A temperature increase of 3.0℃ or more would likely be catastrophic, exceeding the capacity and capability of any risk reduction measures. A report released on 12 October by the Centre for Research on the Epidemiology of Disasters and the United Nations Office for Disaster Reduction illustrates that these prognostications are coming true.

Increase in Disasters

The total number of climate-related disasters - droughts, extreme heat, floods, storms, and wildfires - increased to 5,535 for the 2000-

2019 period, up from 3,009 for the 1980-1999 period. Floods and storms were most common in both periods, with the greatest increases for extreme heat (330 percent) and flood (230 percent) events.

Lives Affected

Almost four billion people were affected by drought, extreme heat, flood, and storm events during the 20 years. There was no global data available for wildfire. Floods (1.65 billion affected), drought (1.43 billion affected), and to a lesser extent storms (727 million affected) affected the most lives. Extreme heat affected 118 million people.

Deaths

The climate-related disasters during the 2000-2019 period resulted in 490,252 deaths. Most deaths were from storms (41 percent), extreme heat (33 percent), and floods (21 percent).

Economic Costs

Climate-related drought, flood, storm, and wildfire disaster events had an economic cost of US$2.26 trillion from 2000-2019. There was no global data available for extreme temperature costs. Storms incurred most of the costs (61 percent, US$1.39 trillion) but floods (29 percent, US$651 billion), and to a lesser extent wildfires (4 percent, US$93 billion) were also costly.

Disaster Reduction Actions

Even with climate action the number and impacts of droughts, extreme heat, floods, storms, and wildfires are likely to increase for many years. The frequency, intensity, and duration of these disasters are up to us. Disaster reduction actions include:

- Enhance communication to help people interpret impending disaster warnings and increase weather forecast accuracy.
- Implement existing, affordable, and effective flood control practices.
- Develop cost-effective measures in poor regions at high risk of recurrent flooding in conjunction with malnutrition programs.
- Deploy life-saving measures for storm impacts like cyclone shelters, wind-resistant buildings, and preservation and restoration of protective mangrove forest and coral reef ecosystems.
- Implement early warning systems and response mechanisms to limit deaths in drought-vulnerable populations.
- Standardize methodologies for collecting comprehensive national data.
- Improve understanding of disaster impacts and analyses to help decision-makers prioritize and target new measures.
- Improve reporting of, and methodologies for, economic losses particularly for lower-income countries.

TOO HOT FOR LIVING

As you heat the planet up, it's just like boiling a pot.

Elon Musk, entrepreneur and provocateur

The Heat is Building

The frequency and duration of excessive heat have increased throughout the world since the 1950s. Future climate projections envision spreading zones of reduced habitability and thousands more deaths annually.

In the United States, most cities are experiencing more days of hazardous heat and warmer nights. In Philadelphia, the average temperature has risen 2°F since the 1950s. Residents this year experienced 40 days of excessive heat, five times more than in 2000.

Phoenix averages about 175 days at or above 90°F and 110 days at or above 100°F. Residents experienced ten excessive heat warnings in 2019 each lasting an average of 3 days. The low temperature stayed above 90°F 16 times during July.

Humans are reaching the limits of their 11 to 15°C mean annual temperature environmental niche. Humans, crops, and livestock have occupied this niche for the last 6,000 years. Parts of Florida are already outside the niche. Unchecked climate warming will push one-third of the global population beyond its limits.

The Cost of Increased Heat

Humans can acclimatize to heat by spending time outside, but not without repercussions. The International Labor Organization reports 1.4 percent of total working hours were lost globally in 1995 owing to heat stress. This is equivalent to thirty-five million full-time jobs. With increasing temperatures, the numbers are projected to rise 2.2 percent by 2030, equivalent to eighty million full-time jobs.

Adaptation options are limited. Humans could migrate to cooler regions of the planet, an action whose geographic, ecological, economic, and political feasibility is unknown.

Or humans could stay in place, relying on air conditioning to survive excessive heat events. The air conditioning option is challenged by the unavailability, affordability, and power outages. For example, 91 percent of indoor heat-associated deaths in Phoenix are related to power shut-off, broken air conditioners, or air conditioners set too low or turned off to save power.

The real cost is measured in loss of life. Excessive heat can cause organ failure. In recent years in the US, about 12,000 people died from heat-associated illnesses. Looking ahead, the number of annual deaths is predicted to increase to 36,000 under moderate climate warming and 97,000 under high climate warming. The estimates include expected adaptation.

Especially vulnerable are communities of color and low income - those least able to afford air conditioning or avoid outdoor work. Older adults are also more vulnerable. Individuals over sixty-five are less able to regulate internal heat and suffer from chronic diseases that exacerbate the effects of heat and dehydration.

Solutions

The ultimate solutions to heat-related deaths and illness are to limit climate change, alleviate poverty, and care for elders. A reduction in greenhouse gas emissions and restoration and protection of carbon sinks will minimize the frequency and duration of hazardous days. Alleviating poverty will enable everyone to have a cool place to avoid the heat. Care for elders will ensure older individuals, especially those alone, will take the necessary mitigation steps and receive treatment before it is too late.

Several US cities are developing solutions to heat-related deaths. Philadelphia began its process in 1993 when a heat event killed at least 118 people. The city focused on the relationships between aging, poverty, and heat determining that the most vulnerable lived in the hottest neighborhoods. Lack of shade, black roofs and paving, and urban heat island effects produced temperatures up to 22°F greater than in other neighborhoods. The coolest places in the city were along the rivers and near large parks.

Philadelphia's action plan encompasses media warnings, tips for how to keep cool, checks on friends and family, cooling centers, halting utility shut-offs, outreach to the home alone and homeless, and block captains. The city estimates they save an average of forty-five lives each year. Chicago has a similar plan.

Phoenix's dangerous hot season extends from May to October, with heat-associated deaths peaking in July, August, and September. The city's cooperative Heat Relief Network includes dispersed cooling and hydration stations, heat refuges, water donation sites, and revised urban designs to encourage shade and cool outdoor corridors. Deaths decreased in 2013 and 2014 but have now risen to new heights.

The recent holistic, multi-sector Heat Action Planning Guide for Neighborhoods of Greater Phoenix describes ways to advocate and educate, improve safety, comfort, and the ability to cope, and build capacity. Many of the actions are like Philadelphia's but with a greater emphasis on urban design.

Excessive Heat Events Guidebook

The Environmental Protection Agency produced an *Excessive Heat Events Guidebook* identifying characteristics that affect health risks and notification and response programs. Elements include continued and enhanced public education, long-term preparation, and response actions. The latter include:

- Encouraging the public to spend time in available air-conditioned buildings.
- Reallocating resources to address critical short-term needs likely to provide significant health benefits.
- Prohibiting the suspension of electric and water services.
- Establishing medical examiner diagnosis standards.
- Conducting a direct assessment of high-risk individuals.
- Increasing emergency medical system capacity.

Recommendations for individuals include:

- Use air conditioning and take cool baths or showers.
- Minimize direct exposure to the sun.
- Stay hydrated.
- Eat light, cool, easy-to-digest food.
- Wear loose-fitting, light-colored clothing.
- Familiarize yourself with the symptoms of heat stress.

A THAWING ARCTIC REQUIRES URGENT ACTION

The Arctic Ocean has lost 95% of its oldest, thickest ice since 1985.

National Oceanic and Atmospheric Association, 2018

Arctic sea ice plays a critical role in reflecting solar energy to space, buying us time to limit climate change. Unfortunately, one-half the surface area and three-quarters of the volume of Arctic ice has been lost over the last few decades. Much of the remaining ice is thin and seasonal, transmitting rather than reflecting solar energy.

Losing sea ice reflectivity equates to 25 years of global carbon dioxide emissions. Climate change is accelerating, and we will reach the 2°C temperature increase that will trigger global, catastrophic heat waves, coastal flooding, species extinctions, and threats to food supplies sooner than originally projected.

The consequences of Arctic ice loss are apparent.

- Arctic surface temperatures continue to rise at twice the rate elsewhere on Earth.
- Snow cover is declining.
- The Greenland Ice Sheet and lake ice are melting, increasing river discharge.

- Ocean primary productivity is as much as 500 percent higher than normal.
- Harmful toxic algal blooms threaten food sources.
- Thawing permafrost is releasing the greenhouse gases carbon dioxide, methane, and nitrous oxide into the atmosphere.

Coast land-fast sea ice is receding, disrupting traditional cultures, and reducing community hazard protection, hunting, and traveling. Habitat for polar bears and other Arctic wildlife is disappearing. Only a few endangered polar bear subpopulations in the deep Arctic are likely to exist by 2100. Caribou and wild reindeer have declined by 50 percent in the last 20 years.

Reducing greenhouse gas emissions will not produce results soon enough to avoid global catastrophe. Action in the Arctic is urgently needed but we have few options. Most hopeful is an initiative by the Arctic Ice Project (formerly Ice 911) to preserve and restore sea ice by applying floating, reflective materials. Early tests are encouraging, and climate modeling suggests that the method can cool the Arctic, rebuild ice area and volume, and reduce the rate of Arctic and global temperature rise.

Another climate change mitigation approach would introduce particles into the atmosphere to reflect solar radiation, mimicking the temperature-reducing effect of volcanic eruptions. The approach is risky, and we do not have enough knowledge to evaluate it yet.

HOW HEALTHY ARE OUR WORLD HERITAGE SITES

Heritage is our legacy from the past, what we live with today, and what we pass on to future generations. Our cultural and natural heritage are both irreplaceable sources of life and inspiration.

United Nations Educational, Scientific, and Cultural Organization

Assessing World Heritage Sites

The United Nations Educational, Scientific, and Cultural Organization (UNESCO) designates cultural, natural, and mixed World Heritage Sites. Currently, there are 1,121 sites, 252 of which protect natural values.

Natural sites are celebrated as the most significant protected areas on Earth, valued for their natural beauty, exceptional natural phenomena, geoheritage, and biodiversity. A 2020 report reveals that the values of World Heritage Sites are deteriorating.

The International Union for Conservation of Nature (IUCN), the official nature advisor for UNESCO, assessed nature value sites in 2014, 2017, and 2020. IUCN's objectives were to track conservation state, recognize well-managed sites, identify pressing issues, and understand and communicate site benefits. Assessment criteria included state and trend, threats, and protection and management effectiveness.

The assessment assigned sites to 1 of 4 categories.

- ***Good*** – Values are currently in good condition and are likely to be maintained for the foreseeable future.
- ***Good With Some Concerns*** – Values are currently in good condition and likely to be maintained if minor additional conservation measures are put in place to address existing concerns.
- ***Significant Concerns*** – Values are considered threatened by several current or potential threats with significant additional conservation measures required to preserve values in the medium to long-term.
- ***Critical*** – Values are severely threatened and require urgent, additional, and large-scale conservation measures or values may be lost.

Assessment Results Overview

In 2020, the 63 percent of the 252 natural values sites were assessed as *Good* or *Good with Some Concerns* (N=111). Thirty percent of the sites had *Significant Concerns* and seven percent were *Critical*. Ninety percent of North America sites, 73 percent of Oceania sites (comprised of Australasia, Melanesia, Micronesia, and Polynesia), 72 percent of Asia sites, and 69 percent of European sites were *Good* or *Good With Some Concerns*.

In both the 2017 and 2020 assessments, most sites were *Good with Some Concerns*. *Significant Concerns* was the second most common assessment finding. Few sites were *Critical*.

Since 2014, the conservation outlook of 17 percent of the sites had changed at least once - eighteen for the better and twenty-five for the worse. Most of the change consisted of deterioration at *Good* sites, which decreased by 15 percent between 2014 and 2020.

Positive developments between 2017 and 2020 included:

1. Wadi Al-Hitan (Whale Valley), Egypt improved from *Good With Some Concerns* to *Good* due to increasingly effective management.
2. Comoé National Park, Côte d'Ivoire improved from *Critical* to *Significant Concern* due to a stabilized political situation.
3. Monarch Butterfly Biosphere Reserve, Mexico improved from *Critical* to *Significant Concern* due to enhanced protection and management efforts.

Conversely, negative developments between 2017 and 2020 included:

- Great Barrier Reef, Australia deteriorated from *Significant Concern* to *Critical* due to climate change.
- Islands and Protected Areas of the Gulf of California, Mexico deteriorated from *Significant Concern* to *Critical* due in part to the imminent extinction of the vaquita - the world's smallest porpoise.

Protection and Management

Protection and management are significant factors in the maintenance of site value. Assessment categories are *Highly Effective, Mostly Effective, Some Concern*, and *Serious Concern*.

Only half of the natural values of World Heritage Sites were assessed as effective in 2020. Ten percent were deemed *Highly Effective* and 40 percent as *Mostly Effective*. Forty-one percent of sites were assessed as *Some Concern* and 9 percent as *Serious Concern*. Sites tended to be more effective for research, legal, and boundary aspects and less effective for sustainable financing and law enforcement.

Oceania is the region with the greatest percentage (95 percent) of effectively managed sites (*Highly Effective* or *Mostly Effective*). North America was a distant second (77 percent). The Arab states (12.5 percent), Mesoamerica and the Caribbean (28 percent), Africa (30 percent), and South America (32 percent) struggled to protect and manage their World Heritage Sites.

Threats

Climate change was assessed as a high or very high threat at 83 of the 252 natural values sites. Existing or potential impacts include increased fire frequency and severity, coral bleaching, severe weather damage, and drought.

Tourism was a significant threat to elements of exceptional beauty, natural phenomena, and geology while invasive alien species were a significant threat to ecological processes, species, and habitats. Other significant threats included dams, water management, hunting, trapping, fishing, fire suppression, and water pollution.

Threats differ regionally in type and severity. For example, Africa, Asia, Mesoamerica, and the Caribbean were threatened by direct resource uses like hunting and fishing. Solid waste, especially plastic pollution of marine areas, was a threat to sites in the Arab states. Livestock grazing was a threat to South American sites.

Improving Site Health

The condition of our natural World Heritage Sites is an indictment of our collective commitment to the protection of the most significant protected areas on Earth. There is no single, simple solution other than to enhance global environmental citizenship.

Climate change is the overarching threat to World Heritage Sites' health. All countries hosting World Heritage Sites should sign on to the Paris Climate Pact, and institutionalize greenhouse gas emission reductions and carbon storage measures. Developed countries responsible for the bulk of emissions should be encouraged to accept responsibility for protecting sites throughout the world.

Efforts to address impacts from tourism, invasive alien species, dams, water management, hunting, trapping, fishing, fire suppression, water pollution, illegal logging, and agricultural encroachment require regional, national, and local collaboration. In some instances, resolving population poverty and food and water security are prerequisites.

Thereafter, governments must support and enforce protection and management efforts. Sustainable financing plans, inclusive of tourism revenue and contributions from businesses and nongovernmental organizations, must be developed under the leadership of local, national, and regional governments.

SAVE THE RAIN FORESTS OF THE SEA

The sea, the great unifier, is man's only hope. Now, as never before, the old phrase has a literal meaning: We are all in the same boat.

Jacques Cousteau, oceanographer, filmmaker, and author

Coral reefs have existed for at least 485 million years. They occupy only 0.1 percent of the world's ocean but provide habitat for 25 percent of all marine species.

Reefs are dying. Between 25 and 50 percent of the world's live coral was lost over the last 30 years. If the trend continues, 70 to 90 percent of all coral reefs will be lost by the middle of the century, and there will be no healthy coral reef ecosystems.

Why Are Coral Reefs Dying?

Direct Human Impacts

Coastal development, dredging, quarrying, and coral harvesting for jewelry and souvenirs have the greatest physical impact on coral reefs. The aquarium trade's over-harvesting of some species reduces biodiversity.

Overfishing alters the reef food web and reduces the number of grazing fish that keep corals clean of algal overgrowth. Destructive practices like blasting and cyanide fishing greatly amplify the damage.

Coastal development, urban stormwater runoff, forestry, and agriculture deposit sediment onto reefs, smothering corals and interfering with feeding, growth, and reproduction. Trash blocks the sunlight needed for symbiotic, photosynthetic zooxanthellae algae that supply the coral with sugars, fats, and oxygen. Other pollution entangles and kills reef organisms, and breaks or damages corals. Degraded plastics and microplastics are consumed by coral, fish, sea turtles, and other reef animals, introducing toxins and blocking digestive tracts.

Nitrogen and phosphorous from agricultural and residential fertilizer, sewage discharges, and animal waste spur algal growth that blocks sunlight and consumes oxygen. Pathogens from inadequately treated sewage, stormwater and runoff from livestock pens can cause disease in corals, especially if they are stressed by other environmental conditions.

Toxic metals, organic chemicals, and pesticides in industrial discharges, sunscreens, urban and agricultural runoff, mining activities, and runoff from landfills affect coral reproduction, growth, and other physiological processes. Herbicides kill coral's symbiotic zooxanthellae, halting photosynthesis and causing starvation. Ultraviolet filters in sunscreens induce zooxanthellae-destroying viral infections, even at extremely low concentrations.

Climate Change

Climate change is the most challenging killer of coral reefs. Rising ocean temperatures stress corals, causing the expulsion of symbiotic zooxanthellae. The corals turn white, a process known as bleaching, become immunocompromised, and eventually starve and die if the stress is not reduced. Even if global surface temperatures are stabilized at 1.5°C above pre-industrial levels, 70 to 90 percent of coral reefs could be lost by the middle of this century.

Acidification is another climate change-related stressor of coral reefs. Oceans absorb carbon dioxide from the atmosphere. The carbon dioxide reacts with seawater to form carbonic acid. Increased carbonic acid increases ocean acidity, reducing the availability of dissolved salts and ions needed by corals to form their calcium carbonate structure. Coral and reef growth slows, and if acidification becomes severe enough, coral skeletons dissolve. Ocean acidity has increased by 30 percent since the beginning of the industrial revolution and is expected to increase by an additional 40 percent above present levels by the end of this century.

Climate-induced sea-level rise, changing weather patterns, and current shifts are also threats to coral reefs. Increased sedimentation smothers coral reef ecosystems. More frequent and intense storms directly damage reefs. Precipitation exacerbates nutrients, pathogens, toxins, and debris in stormwater runoff. Shifts in ocean currents could alter water temperatures, inter-reef connectivity, food supplies, and larval dispersal.

The aggregate impact of direct anthropogenic and climate change stressors decreases reef resilience. Throughout geological time, coral reefs always recovered from warm periods, but each recovery required tens of thousands of years. At no time in history have coral reefs waned so quickly, and recovery thousands of years from now is too late.

Coral Reefs are Too Valuable to Waste

The degradation and loss of coral reefs will depress economies, diminish food security and human health, reduce storm protection, and decrease biodiversity.

Reduced Goods and Services

About 850 million people live within 100 km of the coast and derive benefits from coral reefs. Approximately half a billion people depend on coral reef ecosystems for food and income from fisheries and tourism.

The economic value of coral reefs is estimated at US$375 billion per year. More than ninety-four countries and territories benefit from reef tourism. Australia's Great Barrier Reef provides almost 54,000 full-time jobs. In the United States, coral reefs provide about US$2 billion in flood risk reduction services. Tropical, small island nations are particularly vulnerable to economic losses from damaged or lost reefs.

Impacts on Human Health

Coral reef ecosystems are the source of many medicines. Corals, sponges, sea hares, sea slugs, and other reef animals contain molecules that express anti-inflammatory, viral, tumor, and bacterial effects. Treatments for cancer, Alzheimer's and heart disease, viruses, and inflammation are derived from reef animals. Pharmaceutical companies are 300-400 times more likely to find new drugs in coral reefs than in terrestrial ecosystems.

Exposed Coastal Communities

Coral reefs function as a natural barrier, dissipating 97 percent of a wave's energy and protecting more than two hundred million people from storms, tsunamis, floods, and erosion. Building seawalls for the same level of protection costs $2.5 million per mile.

Less Biodiversity

Coral reefs occur in more than one hundred countries and territories, providing habitat, feeding, spawning, and nursery grounds for over one million species. This is 25 percent of all marine life including 4,000 species of fish. Degraded and lost coral reefs can be expected to impact linked seagrass, mangrove, and mudflat communities, further reducing marine diversity.

Saving Coral Reefs is Possible

Conserving the world's coral reef ecosystems requires multiple actions and the continued development of innovative and scalable solutions. The scale and complexity of coral reef conservation demand stakeholder engagement to ensure actions are effective and sustainable.

The success of every action and solution is dependent on mitigating climate change. Even if we stabilize global surface temperatures at 1.5°C above pre-industrial levels, 70 to 90 percent of coral reefs could be lost by the middle of this century.

Establish Protected Areas

Establishing and expanding protected areas is a demonstrable way to enhance and preserve coral reef health. Protected areas prohibit or restrict fishing, mining, and recreation activities. Protected coral reefs are not immune to climate change, but a recent study suggests they may be more resilient than unprotected reefs.

United Nations Sustainable Development Goal 14 targets conserving at least 10 percent of coastal marine areas. Given the substantial percentage of coral reefs lost or threatened, a higher percentage for reefs is arguably necessary. Protected areas will be most effective when distant from human development and activities unless coupled with a strategy for reducing or eliminating anthropogenic impacts.

Restore Degraded and Lost Reefs

Coral reef restoration takes many forms. At its simplest, the United States National Oceanographic and Atmospheric Administration and the State of Hawaii use a vacuum to remove invasive algae. Sea urchins are then released to eat returning algae.

Nearly 7,000 coral fragments were rescued in Puerto Rico after Hurricane Matthew. The fragments were re-attached to the reef by lodging them in crevices or cementing them to the hard bottom of the reef. Survival rates exceeded 90 percent after three years.

Coral can also be farmed *ex-situ* by collecting and growing naturally produced eggs and sperm, or harvesting coral fragments. The farmed coral may be relocated to *in-situ* farms for additional growth, or attached to existing reefs.

The most recent advancement in coral reef restoration is selecting resilient coral stock for farming and transplanting. Coral tolerance to warmer water has been identified in the Bahamas and Palau.

Promote Awareness, Education, and Advocacy

Awareness of coral reef degradation and die-off is low compared to other environmental crises. Educating stakeholders and converting them into advocates for coral reef health is an essential complement to protection and restoration. Fruitful topics include the relationship between reef health and livelihoods, sustainable fisheries, forming a local advocacy group, and responsible snorkeling and diving.

Manage Near Shore Development and Waste Disposal

The most essential near-shore actions are to halt poorly planned coastal development, eliminate direct damage, and reduce pollution. For existing development, wastewater, and stormwater management can mitigate sedimentation and the discharge of nutrients, pathogens, toxins, trash, and microplastics.

Resorts can adopt best practices to reduce or eliminate fertilizer use, ask visitors to use reef-safe sunscreen, and implement recycling, user-friendly trash collection, water conservation, and water quality monitoring programs.

Agricultural operations in coral reef ecosystem watersheds can also adopt stormwater management measures to reduce or eliminate pollutant discharges. Similarly, destructive feral animals like goats and pigs can be eliminated from areas of the watershed that drain directly to the coastline.

Govern Responsibly

Countries and territories with reefs can establish and implement reef-protecting water quality monitoring and management programs. New and existing programs can regularly re-evaluate their standards, monitoring protocols, and enforcement activities to ensure healthy coral reef conservation is achieved.

The United Nations Conference of the Parties for the Climate Change Agreement can re-evaluate the global warming target. Also, the United Nations Development Programme can provide leadership on poverty, inequality, and exclusion issues that result in reef-destroying behaviors. Multiple United Nations programs can promote international cooperation.

At national and local levels, governments can address unregulated and destructive fishing, anchor damage, and pollution. Success will engender regulation and behavioral change.

Increase and Prioritize Funding

The United Nations Environmental Programme, the International Coral Reef Initiative, the World Wildlife Fund, the Nature Conservancy, the Wildlife Conservation Society, Vulcan Inc., the Ocean Agency, and the Secretariat of the United Nations Convention on Biological Diversity have committed to funding coral reef conservation efforts.

The Global Fund for Coral Reefs was launched on September 16, 2020, at the 75th Session of the United Nations General Assembly. The Fund, a financial instrument blending private and public funding, seeks to raise and invest US$500 million for coral reef health and sustainability over the next 10 years. Actions include empowering local communities and enterprises. Fund contributors include the Paul G. Allen Family Foundation, the Prince Albert II of Monaco Foundation, BNP Paribas, Althelia Funds, the United Nations Development Programme, the United Nations Environmental Programme, and the United Nations Capital Development Fund.

Act

Everyone can help save coral reefs. Read about reef ecosystems, visit an aquarium, plan a responsible snorkeling or diving vacation, and volunteer to help.

WE NEED SHARKS AND RAYS

Sharks are beautiful animals, and if you're lucky enough to see lots of them, that means that you're in a healthy ocean. You should be afraid if you are in the ocean and don't see sharks.

Sylvia Earle, marine biologist, oceanographer, explorer, author, and lecturer

The Benefits of Sharks and Rays

Sharks and rays are critical components of marine ecosystems.

- Sharks reduce climate change impacts by preventing overgrazing of seagrass beds that capture carbon thirty-five times faster than tropical rainforests.
- Large bottom-dwelling rays create microhabitats for invertebrates and uncover food for fish and seabirds.
- Deep diving sharks and rays transfer nutrients from the depths to the surface, promoting the growth of phytoplankton that produces half of the Earth's oxygen.
- Divers pay millions of dollars into local economies each year to view sharks and rays.

- When sharks and rays die, their bodies become food for scavenger communities.

Sharks and Rays Are In Decline

There are more than 2,100 species of sharks and rays in the world's oceans. Sharks and rays, essential to a healthy marine ecosystem, are declining at an alarming rate - 30 percent are threatened with extinction. Up to one hundred million sharks and rays are killed each year for food, trophies, as fisheries bycatch, or out of fear.

Actions to Save Sharks and Rays

Andy Cornish of the World Wildlife Fund offers six actions to save sharks, rays, and the marine systems they support.

Traceability System

US$1 billion in shark and ray products are traded each year, some of which were illegally caught and exported. TRAFFIC - the wildlife trade monitoring network- is developing a traceability system that will expose illegal trafficking.

Distinguishing Shark and Ray Products

The seafood industry needs to encourage shark fisheries to reject unsustainable fisheries, especially of endangered species.

Reduce and Eliminate Bycatch

Non-selective fishing gear needs to be phased out and replaced by gear targeted toward select, commercial species. In the meantime, the use of non-selective gear should be prohibited in shark and ray critical habitats.

CITES

The Convention on International Trade in Endangered Species - which lists twenty-four species of sharks and rays - must accelerate the development of sustainable fisheries. CITES is the only convention that all shark and ray fishing countries are a party. Compliance is mandatory.

Prevent Extinction

Innovative, scalable, conservation measures are needed to prevent the extinction of many shark and ray species while fishery managers implement sustainable policies and fishers develop alternatives to non-selective gear.

Urgency

The conservation community needs to communicate to decision-makers and the public the important role of sharks and rays in healthy marine ecosystems, and the need for immediate action.

SEA SHEPHERD

SQUID TO DIE FOR

This is the largest known case of illegal fishing perpetrated by a single industrial fleet operating in another nation's waters.

Jaeyoon Park, a data scientist from Global Fishing Watch

Soldiers to Fishermen

Each day is an abject lesson about the link between the environment and people. The latest is corpse-filled ghost boats coming ashore in Japan, a complex system of cash-strapped North Koreans, illegal Chinese fishers, squid, and a nonprofit conservation organization co-founded by Google.

Prior to 2017, China paid millions of dollars to North Korea for a license to fish in their waters. But international sanctions imposed on North Korea after ballistic missile launches and nuclear tests forbid foreign fishing. To replace lost revenue, Kim Jong Un pushed the state-owned seafood industry to increase its catch. Soldiers were turned into poorly equipped fishers on small, wooden, boats with limited fuel supplies, and no beds or toilets.

A Chinese Invasion

Meanwhile, a Chinese industrial fleet of hundreds of boats, ten times larger than their Korean counterpart's fleet, illegally returned to North Korean fishing grounds. The evidence comes from a North

Korean fisher who survived a landing in Japan, sightings by South Koreans, and the technological sophistication of Global Fishing Watch, the nonprofit co-founded by Google. Collaborating with an international team of academics, Global Fishing Watch was able to track the Chinese fishing fleet by satellite on their journey from China to North Korean waters.

The Chinese vessels aggressively displace the smaller North Korean boats, forcing them to travel further to fish. Ill-equipped for long voyages far from shore, the Korean fishers are stranded at sea or wash up on Japanese beaches. Autopsies reveal that the men die of starvation, hypothermia, or dehydration. So many fishers are lost at sea that some North Korean villages are referred to as 'Widow Villages.'

A Collapsing Fishery

And the situation gets worse, as marine researchers fear a collapse of the squid population. Pacific flying squid spawn in South Korean waters, swim north to North Korean waters in the spring and return to South Korean waters between July and September. Chinese fishers are harvesting 160,000 tons of squid in North Korean waters, as much as South Korea's and Japan's combined catch in their own waters.

The Chinese are catching the squid before they mature enough to procreate, reducing the population's ability to recover. Once abundant squid stocks are down more than 70 percent, threatening the food supplies and cultures of several countries.

A PATH TO HEALTHY LANDS AND WATERS

In order to carry a positive action we must develop here a positive vision.

The Dalai Lama

A Path Forward

Our present course looks bleak for future generations, but all is not lost. The Intergovernmental Science-Policy Platform on Biodiversity and Ecosystem Services calls for transformative change, a fundamental, system-wide reorganization across paradigms, goals, and values. Noted environmentalist Erle Ellis posits that a better future for nature and people is about motivating people to work together and creating the social processes, institutions, and strategies to make planetary cooperation possible.

The Millennium Assessment evaluated scenarios for just such an outcome, concluding that a globally connected society and actions focused on regional watershed-scale ecosystems offer potential. Less promising are scenarios relying on technology and highly managed ecosystems. A scenario envisioning a regionalized and fragmented world is unlikely to sustain healthy lands and waters or their ecosystem services.

Actions

What actions can we take to achieve transformative environmental change, motivate people to work together, and support promising scenarios? Historically, we have relied on three fundamental practices – regulation, protection, and education. These practices have produced notable successes including recovery of protected species, conservation of critical lands, and increased scientific literacy. Each will continue to be important moving forward.

But as illustrated by the continued decline of our lands and waters regulation, protection, and education alone as currently practiced cannot guarantee the comprehensive healthy lands and waters needed to ensure economic prosperity, health, and well-being. Why not? The regulation of lands and waters requires adequate enforcement and can be weakened or eliminated by shifts in political ideology. Protected lands require significant capital outlay and ongoing maintenance, rarely encompass entire ecosystems, and are susceptible to activities beyond their borders. Education is place, age, and culture-specific, and in the United States and many other countries focused on global competitiveness. Regulation, protection, and education are all ineffective when contrary activities are motivated by greed or base needs for food, water, warmth, rest, security, and safety.

So how do we build on past practices to avoid further declines and reverse historic destruction, degradation, and unsustainable use of our lands and waters? Understanding context is critical. Sustaining healthy lands and waters is complex, requiring science, policy, implementation, and adaptive management in response to dynamic stressors. We face a knowledge burden requiring lengthy education or narrowing expertise, both of which reduce individual innovative capacity. And we have insufficient quality data about threats to healthy lands and waters.

Success demands that we develop the expertise and skills to meet these challenges and to test and implement promising scenarios and augment existing regulation, protection, and education practices. Three areas seem promising.

Collaborate and Innovate

First, we can adapt our formal and individual education to de-emphasize goal of global competitiveness and to favor a goal that prepares students to tackle large-scale challenges through collaboration, creativity, and innovation. A goal enabled by problem-based social and emotional learning. A goal that produces solutionaries - people who develop practical, effective, and visionary solutions that are regenerative, healthy, and just for all.

Heurism

Second, we can emphasize heurism instead of knowledge. For too long our education system has been designed to produce knowledge workers taught to follow a set of established instructions down a single pathway to one conclusion. These left brain leaning knowledge workers score well on IQ tests. But left-brained aptitude only accounts for 4 to 10 percent of career success.

The student of the future will have social, emotional, and technological skills leaving physical, repetitive, and basic cognitive tasks to machines. We must provide people with the capacity to detect patterns and opportunities, synthesize information, craft satisfying narratives, and combine seemingly unrelated ideas into something new. In other words, we must transition from an economy and a society built on the logical, linear, computer-like capabilities of the Information Age to an economy and a society built on the inventive, empathic, big-picture capabilities of the Conceptual Age.

Boundary Crossing

Third, we need to become boundary individuals and boundary organizations. Boundary was originally conceived to describe an organization that mediates between science and policy development. More recent conceptualization envisions organizations that operate in multiple fields, on different scales, and with different stakeholders. I maintain that the concept applies equally to individuals.

Within the context of healthy lands and waters, boundary individuals and organizations facilitate collaboration and the flow of information across the natural and social science disciplines, across science and technology, and ultimately across science, technology, and communities. Not only communities as social groups that reside in a specific locale and share governance, but also groups that share common characteristics or interests such as scientific, political, cultural, and religious communities.

Leadership

The shift from knowledge to heurism, from logical and linear to inventive, empathic, and big picture, from specialization to boundary crossing requires skills associated with leadership. Skills that strengthen character, clarify values and beliefs, further decision-making, and embrace change. Skills that build trust, promote change, create a shared vision, and motivate and enable others. Skills that facilitate effective communication, de-escalate stress, manage group dynamics, and identify and analyze stakeholders. Skills that promote creativity and innovation. And ultimately, skills that build communities of practice.

GROW TRUST IN SCIENCE BY GROWING SCIENTISTS

Science is a way of thinking much more than a body of knowledge.

Carl Sagan, astronomer, planetary scientist, author, and science communicator

Only 29 percent of American adults have a great deal of confidence that scientists are acting in the best interests of the public according to a 2022 Pew Research Center Report. This is a 10 percent decrease in trust since 2020. If you think this is a partisan issue, you're wrong. Democrats are only slightly more likely to express confidence in scientists than Republicans.

Science is the process of objectively establishing facts through testing and experimentation. The process is designed to produce relatively unbiased and accurate results that serve as the basis for decision-making. The scientific process has been tested over time and shown to be reliable.

So how do scientists rebuild trust with the public? Is it doing a better job ensuring the public understands science? Or do scientists commit to improved ethics and greater transparency? The answer is likely both.

Increasing the public's lack of trust requires that scientists take a long, hard look in the mirror. Blaming the public won't change opinions. Here are eight recommendations for growing trust in science by growing scientists.

1. Scientists are Human

Scientists apply the scientific method to ensure that conclusions are less biased than intuition, observation, anecdote, or opinion. Yet scientists are still human, wrestling with an imperfect brain and a propensity for bias influenced by past knowledge and culture. Scientists should acknowledgeable their humanity while striving for objectivity, rigidly apply the scientific method, and encourage others to independently examine their conclusions.

2. Acknowledge That Science is Not Immutable

Scientists are their own worst enemies when they present themselves and their science as immutable. Yes, anything else could weaken public confidence but much less than if our seemingly immutable science is inevitably mutable. Be honest with the public. Science improves with more study and no one study is definitive. The value of science is a community of scientists continually applying the scientific method to relevant problems. Consensus, not uniformity, is the basis for science-based decision-making. Inevitably, both the science and the decision-making will change.

3. Integrity is Paramount

The public will accept mistakes and mutable science but not dishonesty. The scientific method does not begin with a preconceived result that biases experimental design and accommodates a friendly interpretation of the data. The goal is not being right or getting published. And don't prevent others' work from public viewing simply because it contradicts your findings. The goal is to inform science-based decision-making, and in the process make the world better.

4. Diversity and Equity Strengthen Science

Applying the scientific method requires challenging your thinking, and the best science happens when informed by diverse perspectives and subjected to constructive challenges. Build teams across discipline boundaries, learning backgrounds, genders, and cultures. Accept that everyone's thoughts are valuable, and make expressing those thoughts safe from retaliation.

5. Improve Access to Scientific Research

There was a time when scientific research was primarily the domain of educational institutions with libraries that subscribed to expensive scientific journals. Now, credible science comes from private companies, nonprofit organizations, government agencies, and individuals. Not everyone can afford subscriptions and the fee for individual papers. The scientific community needs to continue improving access to scientific work to facilitate diversity, equity, and inclusion, reduce the burden of knowledge, activate collective intelligence, convey ownership to stakeholders, and ultimately produce the best science to inform decision-making.

6. Communicate to be Understood

Science communication is getting better, but is still hampered by jargon and complex grammar and syntax. Yes, jargon and complex grammar and syntax suggest intelligence and facilitate understanding among scientists. But real intelligence is the ability to be understood by critical audiences. Take a writing class or two, read books by good science communicators, and listen to David Attenborough. Learn how to be informative and persuasive. Be smart in what you say but be sure that everyone understands you.

7. **Science is One Piece of a Complex System**

Scientists and science are not always the final word. Science is part of a complex scientific, social, economic, and political system. Scientists must engage and navigate all aspects of the system, either alone or through collaboration, to make science meaningful and trustworthy.

8. **The Public is Counting on Science and Scientists**

Climate change and biodiversity loss are overarching challenges to human well-being and prosperity. The public needs science and scientists more than ever. Scientists need to reintroduce themselves to the public and rebuild trusting relationships. Humility, respect, and service to others are the core of a new public-science relationship. Re-examine your reasons for becoming a scientist and what you hoped to accomplish. Hopefully one of your reasons is well-being and prosperity for all.

DO WE NEED A BIODIVERSITY TARGET?

Look closely at nature. Every species is a masterpiece adapted to the particular environment in which it has survived. Who are we to destroy or even diminish biodiversity?

E.O. Wilson, biologist, naturalist, and writer

A recent scientific publication proposed an annual extinction target of well below 20 species for the next one hundred years. The target is a proffered as a solution for continued biodiversity loss, which the authors attribute to the lack of mainstreaming biodiversity into public policy and limited success raising the profile of biodiversity. Certainly, scientists and policymakers need to do a better job conserving biodiversity. I commend the authors for their leadership, creativity, and provocativeness.

But is a biodiversity target necessary? Extinctions are occurring every year, and a target could arguably focus attention and bring order to the process. But order and process open the door to discussions that could exacerbate biodiversity loss. Any target greater than zero implies that extinction is acceptable. It undermines the argument that biodiversity is critical because all life is interdependent. That removing one species affects all species, including humans.

The phrase well below 20 invites challenge and bargaining. Are five extinctions acceptable? How about 10 or 15? What is the most species that can go extinct each year without impact? Can extinctions be averaged across years? Do extinctions in one country preclude extinctions in other countries?

Do the extinctions happen in a black box i.e., we absolve ourselves of responsibility for any single extinction if the total is well below 20? If we take responsibility for individual extinctions, what criteria do we apply? No scientist understands which species is expendable in the short-term, let alone the long-term. How will scientists reach consensus? Or are extinction choices the prevue of national policymakers?

How can scientists provide knowledgeable, accurate, and persuasive advice to policymakers and the public? What do the policymakers and public do when another scientist inevitably offers a contrary opinion?

How will science balance local, regional, and global needs? How will science balance short-term versus long-term needs? Who will accept responsibility when a scientist's well-meaning advice results in an unforeseen ripple through an ecosystem, further reducing biodiversity and harming communities.

Unlike climate change, an extinction does not provide opportunity for restoration or reversal. Extinction is forever unless we modify our ethics to accept a Jurassic Park approach.

Biodiversity is a complex assemblage defying our ability to accurately predict the consequences of removing a single species. At a time when biodiversity is declining and extinctions are accelerating, accepting any extinction signals a lack of urgency. The target should be zero and we should measure progress towards that goal.

In support of a zero extinction target, the scientific and conservation communities should invest in actions to address mainstreaming biodiversity into public policy and raise its profile with the public.

1. Build a coalition of scientists, communication specialists, and marketers committed to making biodiversity accessible to policymakers and the public.
2. Acknowledge that the concept of biodiversity is relatively abstract compared to killing elephants for their tusks, rhinos for their horns, or the dwindling number of wild pandas in China.
3. Identify a suite of specific examples where changes in biodiversity impacted ecosystem condition, function, and value.
4. Develop stories around the aforementioned examples, relevant to specific audiences, which communicate the problem, why it matters, solutions, and benefits.
5. Be persuasive, using motivational interviewing, Cialdini's 6 principles of persuasion, and what we have learned from copywriting and fundraising.
6. Develop and implement hands-on education programs that help others discover the relevance of biodiversity.

Biodiversity is a critical metric for evaluating the health of our lands and waters. Accepting any extinctions weakens biodiversity's significance and triggers a long, unhelpful debate about implementation, measurement, and enforcement. Time would be better spent helping policymakers and the public discover the importance of biodiversity for themselves.

SUSTAINABLE INVESTMENTS ARE GOOD FOR EVERYONE

"Sustainability, ensuring the future of life on Earth, is an infinite game, the endless expression of generosity on behalf of all."

Paul Hawken, entrepreneur, author, and environmental activist

Corporate social responsibility is a hard-edged business decision. Not because it is a nice thing to do or because people are forcing us to do it... because it is good for our business

Niall Fitzerald, former CEO of Unilever

What is an ESG Investment?

Environmental, Social, and Governance (ESG) investments seek positive economic returns and long-term, positive impact on the environment, society, and business performance.

How do ESG Investments Perform?

A report by the Inter-American Development Bank concluded that ESG investments perform as well as or better than conventional investments. Most academic meta-analyses find that ESG investments to perform as well or better than conventional investments. The Inter-American Development Bank's study of Chilean sovereign and pension funds found that ESG investments would have, in many instances, outperformed the country's conventional funds.

In addition, ESG investments help governments invest in companies that benefit communities and avoid the costs of unsustainable practices.

Are Asset Management Firms Supporting ESG Investments?

More than US$20 trillion, a quarter of all professionally managed assets worldwide, follow ESG strategies. Twenty-five hundred signatories representing US$80 trillion pledged to follow the United Nation's Principles of Responsible Investment.

Principle 1: We will incorporate ESG issues into investment analysis and decision-making processes.

Principle 2: We will be active owners and incorporate ESG issues into our ownership policies and practices.

Principle 3: We will seek appropriate disclosure on ESG issues by the entities in which we invest.

Principle 4: We will promote acceptance and implementation of the Principles within the investment industry.

Principle 5: We will work together to enhance our effectiveness in implementing the Principles.

Principle 6: We will each report on our activities and progress towards implementing the Principles.

Despite good performance, ESG investing could be improved if the investment community agreed on standardized metrics, a transparent reporting system, and a standardized third-party rating system.

AWARD-WINNING PAYMENT FOR ENVIRONMENTAL SERVICES PROGRAM

Economic Valuation and Payment for Environmental Services - Recognizing Nature's Value or Pricing Nature's Destruction?

Jutta Kill, Heinrich Böll Stiftung Foundation

Payment for Environmental Services

Payments for environmental services (PES) are payments to landowners who have agreed to take certain actions to manage their land or watersheds to provide an ecological service. PES is a market-based mechanism that recognizes the important role of the environment in well-being and prosperity.

Costa Rica's Award Winning PES Program

The United Nations selected Costa Rica's Program for Payment of Environmental Services as a 2020 Climate Action Award Winner. The awards are given each year by the United Nations Climate Change Momentum for Change initiative to showcase practical and replicable examples of what people are doing to address climate change.

Costa Rica's Program for Payment of Environmental Services is a financing mechanism for managing, conserving, and sustaining forest resources and biodiversity. The Program for Payment of Environmental Services is built on four pillars.

Pillar 1. Institutionalization

The PES Program facilitates and promotes the inclusion of multiple entities with interests in the development of Costa Rica's forestry sector. Entities include the National System of Conservation Areas, the National Forestry Financing Fund, the National Forestry Office, the Forest Regents, the College of Agronomists, cooperatives, agricultural associations, and non-governmental organizations. The Mesoamerican Biological Corridor initiative also has a voice.

Pillar 2. Legal

The groundwork for the PES Program was laid in the 1990s with legislation favoring the conservation and protection of natural resources. The commitment was subsequently strengthened with Costa Rica's commitments to several global initiatives (e.g., Rio Declaration on Environment and Development, Agenda 21, Climate Change Conventions). New laws were developed to support the PES Program including the right of all inhabitants to the enjoyment of a healthy and balanced environment. The National Forest Financing Fund of the Ministry of Environment and Energy administers the PES Program.

Pillar 3. Financing

The PES Program is sustained by a 3.5 percent tax on fuels. Other funding sources are contributions from the government, donations or credits from national and international organizations, credits from the National Forestry Financing Fund, and agreements with the private sector.

Pillar 4. Monitoring and Evaluation

The National Forestry Financing Fund designed a monitoring and evaluation scheme to ensure that investments are efficient and support providers of environmental services.

PES Activities

Forest Protection

The forest protection objective is to conserve and protect natural resources while sustaining environmental services. Protected, restored, or naturally succeeding forests are characterized by mature trees of different ages, species, and sizes.

Reforestation

The reforestation objective is to establish forest plantations in locations with high productive potential for wood products while contributing to other environmental services. The plantations are cultivated from one or more forest species, with priority to projects that use improved genetic material.

Natural Regeneration

The natural regeneration objective is to conserve and protect natural resources in de-vegetated sites to facilitate recovery and contribute to environmental services.

Agroforestry Systems in Coffee

The agroforestry systems in coffee objective is to establish agricultural systems interspersed with coffee crops. Agricultural production is a polyculture of tree species, coffee, and other perennial crops.

Mixed Systems

The mixed systems objective is to establish agricultural systems that include trees and forest regeneration on farms of less than 10 hectares. The projects establish forest tree species integrated with annual crops.

Protection of Water Resources

The protection of the water resources objective is to conserve and protect water resources within basins where the environmental service is generated. Projects are established in forest areas within aquifer recharge zones of importance for sustaining water regimes and protecting water resources.

Reforestation With Native Species in Extinction

The reforestation with native species in extinction objective is to establish forest plantations with threatened and endangered species and contribute to other environmental services. Reforestation includes habitat for one or more listed species.

Agroforestry Systems

The agroforestry objective is to establish agricultural systems which include trees interspersed with crops or pastures that produce wood, firewood, forage, fruits, medicines, and other environmental services. The tree species include those in danger of extinction.

Agroforestry Systems with Endangered Species

The agroforestry systems with endangered species objective is to establish agricultural systems of trees interspersed with crops or pastures that support threatened or endangered species and contribute to other environmental services. The trees provide products and services, while promoting the ecological importance of listed species.

Post-harvest Protection

The post-harvest protection objective is to protect natural resources in final harvest areas in accordance with management plans approved by the State Forestry Administration.

PES Achievements

Launched in 1996, the positive impacts of the PES Program include:

- Benefits to more than 18,000 families and 100,000 Indigenous people.
- US$524 million invested.
- 1.3 million hectares under contract.
- A reduced rate of deforestation.
- Recovering forest cover and degraded lands.
- Reduced illegal logging.
- Promotion of forest industry and non-traditional production and exports.
- Contributions to rural development and reduced poverty.
- Enhanced compliance with global environmental goals.

The National Forest Financing Fund receives delegations from countries throughout the world interested in the PES Program. Morocco, Peru, and the Dominican Republic are developing or implementing similar initiatives.

CHINESE LESSONS

When one teaches, two learn.

Robert Heinlein, science fiction author, aeronautical engineer, and naval officer

The Dongtan Wetland Park

About 15 years ago, as Executive Director for Community Watershed Fund, I had the good fortune to provide the ecological design for the Dongtan Wetland Park in Shanghai, China. Through enhancement of a degraded coastal wetland, the Dongtan Wetland Park was designed to provide critical habitat for migratory birds and other wildlife, and a reintroduction site for the highly endangered Chinese alligator. A pathway would meander throughout the Park's varied habitats, affording visitors the opportunity to view indigenous plants and wildlife. An education and research center would proffer information on regional natural history, and serves as a base for scientific study.

Most interestingly, the Wetland Park was intended to be is only the first piece of the 6.5 km^2 Dongtan International Wetland Park, whose planning objective is *man and nature living in harmony – a perfect eco-friendly environment constructed by man but that seems created by nature.* The Park is in turn part of the Chongming Ecological Demonstration Project, a progressive and ambitious planned community of residential, agricultural, commercial, and industrial uses integrated with the natural environment.

Dongtan Was Just the Beginning

The scope and vision of the Dongtan/Chongming project are not unique in China. Community Watershed Fund had the opportunity to meet and advise equally ambitious wetland park proponents in Nanjing, Hangzhou, and Wuxi. Nanjing was to be the home of the Qinhuai River Wetland Park, which would integrate the environment, Qinhuai culture, and residential and commercial development. Riverine and wetland habitats would be enhanced, restored, and created to attract fish and wildlife. Access and visitation facilities would accommodate domestic and international visitors, school groups, and scientists.

In Hangzhou, the Xixi Wetland Park would encompass a network of rivers, wetlands, fishponds, and small islands. Wetlands would be restored, cultural protection zones established, and water quality improved. Boats would convey visitors throughout the Park's waterways with stops at temples, fishponds, and agricultural fields. An existing farmhouse restaurant offers visitors an opportunity to dine on local fish and produce.

The Changguang Stream Wetland Park in Wuxi would integrate the environment, Wuxi culture, and development. Riverine and wetland habitats would be enhanced, restored, and created to improve water quality, attract fish and wildlife, and provide recreation. A new development zone, with a university as the centerpiece, was planned to complement the Park.

A view of landscape as wealth predominates in most cultures, and was apparent in the Chinese attitude toward land use. However, the

Chinese view extended beyond simply couching landscape value in terms of dollar per square foot or number of units per acre. As exemplified by the wetland parks in Shanghai, Nanjing, Hangzhou, and Wuxi, the Chinese implicitly and explicitly recognized the inherent value of the natural landscape. The parks were designed to capture ecosystem services, without which the local municipalities would be required to implement expensive infrastructure improvements.

Another aspect of Chinese cultural influence on land use was recognition that quality of life is integrally linked to a healthy habitat. A synergy was envisioned in which people are integrated with, and become part of, a managed nature. In practice, and as exemplified by the projects in Shanghai, Nanjing, Hangzhou, and Wuxi, a systems management approach to planning was adopted. Consequently, the projects address not only economic sustainability, but also ecologic sustainability and the provision of recreational and cultural experiences for local, regional, and international visitors.

I extracted four simple lessons from the Chinese attitude and approach to land use exemplified by the projects in Shanghai, Nanjing, Hangzhou, and Wuxi.

Assets Not Liabilities

Landscapes provide valuable ecological services and are not obstacles to be overcome. Wetlands are especially valuable assets, used directly for recreation and food and indirectly for flood control, storm protection, and water quality renovation.

Now and Forever

The proponents of each Chinese park I spoke with emphasized their desire for sustainability – both ecological and economic. The design to which Community Watershed Fund contributed, and the conceptual designs we discussed, sought to minimize capital, operation, and maintenance costs by working with the land, using adapted plant species, and minimizing infrastructure. Moreover, the parks were expected to generate revenue.

Integrate

Each of the Chinese parks is locally and regionally integrated, i.e., viewed as one component of a much broader sustainable development plan. Most significantly, each of the parks Community Watershed Fund discussed was located proximal to a population a center, which of course facilitated realization of ecosystem services.

Think Big

Bigger is better when it comes to using ecosystem services. The large Chinese wetland parks – 9 to 24 km^2 – can provide more ecosystems services, and at a higher level, than smaller parks. Consequently, conventional infrastructure can be avoided or minimized. The economy of scale also reduces operation and maintenance costs.

Confucius said that we learn wisdom by three methods – 1) by experience, which is the bitterest, 2) by imitation, which is easiest, and 3) by reflection, which is noblest. No culture is exempt from learning, and the best learn from all three methods. The Chinese have gone to great lengths to learn from their experiences, and from those of the U.S. The Chinese, when appropriate, have imitated U.S. approaches, designs, and techniques for using land. But it is upon reflection that the Chinese have wedded watershed management and regional planning to produce sustainable development centered on wetland parks. On this issue, it is time that the U.S. imitates the Chinese.

RESTORING THE SOUTH FLORIDA ECOSYSTEM

... the Everglades ecological system not only contributes to South Florida's water supply, flood control, and recreation, but serves as the habitat for diverse species of wildlife and plant life. The system is unique in the world and one of Florida's greatest treasures.

Everglades Forever Act, 1994.

Florida Statehood and Cropland

Prior to Florida statehood in 1845, the Everglades was perceived by many as a hot, humid environment rendered nearly impenetrable by saw-edged grasses and copious mosquitoes. Florida had a population of about 50,000 people, 90 percent of who lived north of Gainesville. The state's only resource was land, including vast undeveloped and unused tracts south of Lake Okeechobee known as the Everglades. Consequently, Buckingham Smith, attorney, politician, and diplomat, was dispatched to assess the economic value of the Everglades. His 1848 report to Robert J. Walker, Secretary of the Treasury, detailed an ambitious scheme for lowering Lake Okeechobee, draining surplus water to tidewater, and using the drained lands for crop production.

Florida pursued Mr. Smith's recommendation to convert the northern Everglades to cropland. Drainage began in the 1880s and was largely completed by 1920. The U.S. Department of Agriculture established a sugar cane breeding station near Canal Point in 1920, and in 1921 the Florida Legislature established an agricultural experiment station in Belle Glade. Despite these efforts, agriculture was limited by repeated floods and by killer hurricanes in 1926 and 1928. Flooding was alleviated by construction of the Hoover Dike around the south shore of Lake Okeechobee in 1930. Today, south Florida agriculture and agriculture-related industries contribute about $8 billion to Florida's annual gross state product, and provide about 135,000 jobs.

Everglades National Park

The state and federal governments were committed to preservation of the southern Everglades concurrent with agricultural development in the northern Everglades. Four-thousand-acre Royal Palm State Park, later to become the nucleus for Everglades National Park, was established in 1916. The Director of the National Park Service declared in 1923 that an untouched example of the Everglades should be established as a national park. In 1929, the Florida legislature authorized the Tropical Everglades National Park Commission, and the U.S. Congress authorized a feasibility study for a national park in south Florida. The park was authorized in 1934. Little progress was made to establish the park until 1946 when the Florida Legislature appropriated $2 million for the purchase of private lands. Everglades National Park was dedicated in 1947. Four hundred sixty thousand acres at inception, the Park now encompasses more than1.5 million acres.

Central and South Florida Project

Settlement of the lower east coast began in earnest in the early 20th century with extension of the railroad to Miami, but was slowed by hurricanes and the Great Depression. Reinvigorated in the 1930s, settlement was slowed again by heavy rains and hurricanes in the 1940s. The latter prompted the Central and South Florida Project, a joint federal and state initiative of the 1950s to provide flood control and water supply to urban areas.

Levees were constructed between the Everglades and the lower east coast, and canals were deepened to increase drainage. Additional levees were constructed in the 1960s to delimit the Water Conservation Areas – impoundments designed to provide flood protection and to ensure adequate water supplies for Everglades National Park, agriculture, and the burgeoning population. As a result of these efforts, more five million people live in south Florida today, and another 20 to 40 million visit each year.

Comprehensive Everglades Restoration Plan

The Central and South Florida Project was successful at servicing urban and agricultural areas, but was markedly unsuccessful at preserving Everglades National Park. Numerous minor modifications to the Project over the next 30 years attempted to improve water delivery to the Park. Failing to restore Park hydroperiod, the 1992 Water Resources Development Act authorized a re-study of the Central and South Florida Project. The re-study resulted in the two thousand Comprehensive Everglades Restoration Plan - an ambitious, decades-long effort to reconcile restoration and protection of valuable natural resources with the water needs of south Florida residents.

The Everglades Forever Act

Concurrent with the Central and South Florida Project re-study, Florida was responding to a lawsuit filed by the federal government alleging its failure to adequately protect the resources of Everglades National Park and the Arthur R. Marshall Loxahatchee National Wildlife Refuge. The resulting Settlement Agreement and Statement of Principles committed federal, state, and certain agricultural interests to restoring and maintaining water quality and quantity to the Park and Refuge. The 1994 Everglades Forever Act outlines Florida's approach to restoring and protecting the Park and Refuge, and to other parts of the Everglades Protection Area.

Managing the Ecosystem

An historical perspective illustrates that our best intentions at sustainable development in south Florida failed to protect its natural assets. We learned too late that altering one part of the south Florida ecosystem affects other parts of the system. Runoff from urban and agricultural areas, flood storage and water supply, and the importation of exotic plants had unintended impacts on Everglades National Park, the Arthur R. Marshall Loxahatchee National Wildlife Refuge, and the Water Conservation Areas.

Restoring, preserving, and protecting the Florida Everglades requires re-establishing natural hydroperiods, treating urban and agricultural runoff, storing water during the dry season, controlling exotic plants and wildlife, and preventing further encroachment by development into valuable natural habitat. The Comprehensive Everglades Restoration Plan and the Everglades Forever Act are forward-looking instruments for accomplishing these tasks. Together, the Plan and the

Act constitute an honest, practical, and equitable approach to restoring, preserving, and protecting the Everglades and the south Florida ecosystem. Both explicitly recognize that restoring, preserving, and protecting the Everglades will require managing the entire south Florida ecosystem.

THE CRIME OF ECOCIDE

The genocide will not necessarily take the form of war, or death camps. Most likely it will take the form of ecocide, in which landscapes are devastated and the populations that live there slowly starve or turn upon each other savagely because there isn't enough food or water to go around.

Arundhati Roy, best-selling author and political and environmental activist

In France, a Citizens Climate Assembly, the Minister for Ecological Transition, and President Macron signaled their support for a new ecocide law at the International Criminal Court. The proposed law is support by small island states, the Swedish workers movement, the Olof Palme International Centre, and Pope Francis.

Ecocide is defined as *any action which has caused serious ecological damage by participating in the manifest and significant neglect of planetary limits, committed with knowledge of the consequences which were to result there from, and which could not be ignored*. In simple terms, ecocide is the destruction of the natural environment by deliberate or negligent human action. The ecocide initiative is committed to guaranteeing the livability of the planet by holding those responsible for ecosystem degradation and destruction legally and financially responsible.

The proposed law is enabled by the concept of planetary boundaries developed by the Stockholm Resilience Institute. The boundaries were influential in the drafting of the UN's Sustainable Development Goals.

Planetary boundary limits are available for nine processes.

1. Climate change.
2. Loss of biosphere integrity (i.e., biodiversity).
3. Nitrogen and phosphorus flows to the biosphere and oceans.
4. Land system change.
5. Ocean acidification.
6. Freshwater consumption and the global hydrological cycle.
7. Stratospheric ozone depletion.
8. Atmosphere aerosol loading.
9. Chemical pollution and the release of novel entities.

Of the nine boundaries, the researchers report that biosphere integrity, climate change, the flow of nitrogen and phosphorus to the biosphere and oceans, land system change, and chemical pollution and the release of novel entities have already been exceeded.

The Catholic Church's support is part of a movement to include a definition of ecological sins in its official teachings. Ecological sins are viewed as sins against the common home. Pope Francis said that ecological sins can be considered ecocide, and called on the international community to recognize ecocide as a fifth category of crime against peace.

The impetus for a crime of ecocide is the exploitation of shared ecosystem service-providing natural resources to the benefit of some and to the detriment of others. Potential examples of ecocide include:

- The widespread application of pesticides and defoliants during the Vietnam war, destroying up to two million hectares of agricultural lands and forests and causing human chronic and acute health problems.
- The Alberta Tar Sands project resulting in the loss of boreal forest and peat bogs the size of England and the extraction of dirty oil exacerbating climate change.
- The diversion of rivers and the disappearance of the Aral Sea, changing the regional climate and resulting in the disappearance of flora and fauna.
- The killing of millions of bees by neonicotinoid pesticides, precluding pollination of natural and agricultural plants and killing other wildlife.
- Air pollution, including greenhouse gas emissions, which threaten the well-being and prosperity of people throughout the world.

Based on Allan, B. (2008). Knowledge creation within a community of practice

COMMUNITY-BASED CONSERVATION

We need others to complement and develop our own expertise.

Etienne Wenger, educational theorist and practitioner

What is Community-based Conservation?

Community-based conservation is people with a shared interest who come together to solve problems, improve skills, and learn from each other. Communities of practice are collaborative, supported by expertise, focused on a shared goal, and sustained over time.

Members of the community share best practices and explore new ways of doing things. Rooted in constructivism, the emphasis is on learning through complex problem analysis. The collective knowledge of the group advances while advancing the individual member's knowledge.

Community-based conservation can take many forms. Here are three of the most common.

Peer-to-Peer Consults

A peer is equal to you in abilities, qualifications, age, background, or social status. A peer-to-peer consult is a natural, safe, and focused mechanism for you or your organization to connect with other practitioners, learn from their experiences, and develop new ideas. Peer-to-peer consults can be applied within a single organization or among independent professionals who share your practice area.

Four to five peer consultants guide a conversation that helps a case presenter address a problem. One of the consultants facilitates, one keeps track of time, and another takes notes. The consultation takes one hour and emphasizes gathering information and analyzing the problem. The process has six steps.

1. *Problem Presentation* - The case presenter briefly and concisely presents a problem-related question to their peers and provides pertinent information including, stakeholders, stakeholder interests, and what actions they have taken or plan to take.
2. *Problem Clarification* - The peer consultants ask critical questions to clarify the problem. Questions address information presented and information not presented but deemed important.
3. *Problem Analysis* - The presenter moves outside the group and listens quietly while the peer consultants analyze the problem.
4. *Possible Actions* - The presenter remains on the balcony while the peer consultants discuss possible actions to resolve the problem.
5. *Presenter Reflection* – The presenter returns to the group and shares their experience on the consult.
6. *Presenter and Peer Consultants Debrief* – The presenter and peer consultants reflect on how they contributed to the problem-solving process. In the spirit of appreciative

inquiry, everyone should name two aspects they did well and one aspect they could improve.

Enhancing Your Peer-to-Peer Consult

Nonverbal communication - voice intonation, pace, and amplitude, and gestures - can deepen consultant understanding of the presenter's meaning and objectivity. Other clues will help you better understand the presenter's problem and the presenter's level of objectivity.

The presenter framed their problem for the peer consultants. The framing helps the presenter simplify and make meaning out of the situation, but that framing can introduce subjectivity and result in adherence to a story that inhibits problem understanding.

Common stories include being a hero fighting the problem, us versus them, believing they are the only one who knows the truth. A presenter might fashion a problem so large it is unsolvable. Or the presenter might choose to be the objective mediator rather than a stakeholder or be missing from the story. All of these are clues to a presenter's blind spots.

Partner Meetings

A partner is someone who shares or is associated with you and your organization in some action or endeavor. Partners share information or funds, collaborate, and serve as influential supporters and important advisors. The relationship is mutually beneficial and enduring operating most often at the organization's strategic level.

Partners may or may not be entirely aligned in their goals but have committed to an honest, productive relationship that serves the interests of all parties to the extent possible. Partner meetings Create or enhance partnerships, seek input on program direction, promote coordination among partners, and encourage partner investment.

Partner meetings are a process, not a single event. Their structure and process reflect the goals of the partners.

Hosting Communities

Hosting draws on the collective intelligence of communities while creating advocates and identifying people willing to help with implementation. Leaders as hosts promote meaningful conversations among people to gather new insights for action. As a host you are neither a moderator nor a conventional facilitator. Instead, you guide the deep and meaningful conversations of others, and harvest their collective intelligence. The benefits of hosting include better decision-making, capacity-building, a more rapid response to problems and change, and participants empowered and committed to implementation.

Hosting is used to analyze issues, resolve conflicts, and develop strategy. It begins with a clear purpose, selecting and supporting a hosting team, identifying a powerful question, designing a compelling invitation, and preparing the meeting space and the participants. Participants can include anyone with a stake in the issue.

The hosting team decides which hosting method is most likely to create a comfortable environment and produce meaningful conversation. During the meeting, the team guides the process and supports the participants in its quest to collectively learn and discover, then harvesting what the group produces.

Circle

Circle is a method unto itself and a component of other methods. As a method, circle is effective for creating collaborative dialogue for groups of 5 to 25. The circle shape enables eye-to-eye contact, signals equality, and creates a dynamic learning opportunity.

Circle fosters a sense of partnership where everyone knows they are heard. Circle gives people a way to influence a conversation, empowers the participants, and enhances group cohesiveness and effectiveness. Circle leadership is shared and lightly held. Roles include:

- *Host* - A host establishes the circle's intention by developing a powerful question, defining expected outcomes, determining participants, setting an agenda, issuing the invitations, preparing the space, and helping define the scope of the conversation. For multiple meetings, the host position rotates among participants as they become familiar with the process.
- *Guardian* -A guardian works with the host to keep the circle centered and intentional. Notably, the guardian aids self-governance and maintaining group intention by observing process. They call a pause in the conversation using a gentle noisemaker or other device to refocus, note a shift, or capture a decision. Any participant can request a pause.
- *Scribe* - A scribe, typically a circle participant, records insights, decisions, and actions. The scribe can be a single individual or rotated throughout the circle. The recording can be written notes or graphics and is made available to participants after the circle.
- *Participants* -Circle participants accept shared responsibility for the quality of the experience and agree to focus on group cohesion and productivity rather than personal agendas. Participants speak with intention and relevance, listen with attention, and tend to the well-being of the group.

World Café

World café is an expansion of circle able to accommodate hundreds of people. It is well suited to determining what a community is thinking and feeling. Groups of 4 to 6 participants gather around tables to discuss a matter of importance. After each conversation, a host remains at the table while the others disperse to new tables. This enables the threads of the various conversations to be woven together. Participants gain a sense of the collective discoveries.

World café has seven design principles.

1. Set the context by establishing the purpose and parameters.
2. Create hospitable space that feels safe and inviting.
3. Explore questions that matter so that knowledge emerges in response to your questions.
4. Encourage everyone's contribution to optimize learning and collective intelligence.
5. Connect diverse perspectives to increase the possibility of new insights.
6. Listen together for patterns and insights to understand central ideas.
7. Share collective discoveries to harvest the collective intelligence of the participants and make it available to everyone.

Craft Compelling Questions for Circle and World Cafe

Compelling questions attract energy, focus attention, invite inquiry, and surface new ideas. They are simple, clear, thought-provoking, and open new possibilities. They invite people to communicate new ideas.

Compelling questions speak personally, share stories, express hope, address uncertainties and complexities, and expand people's range of experience. You can create compelling questions by focusing on three dimensions – why questions, scope, and unbiased - each helping people change their thinking and inspiring a sense of possibility that encourages action.

LOOK TO INDIGENOUS PEOPLES TO SUSTAIN THE PLANET

Indigenous students learn at an early age how to preserve the value of their lands and waters.

United Nations Under-Secretary-General and United Nations Environmental Program Executive Director Achim Steiner.

Indigenous Peoples and Their Lands

Indigenous peoples have an historic, pre-colonization association with an area, distinct social, economic, and political systems, and are strongly connected to their lands. They own or occupy 25 percent of the Earth's surface, including lands that intersect with 40 to 50 percent of all terrestrial protected areas. The percentage is even greater in the Americas, reaching 90 percent in Central America.

More than 370 million Indigenous people and two billion others rely on products and services from Indigenous and other communal lands. Indigenous lands generate billions to trillions of dollars each year in ecosystems services, including carbon sequestration and water purification. Indigenous lands store 24 percent of the world's above-ground carbon and harbor 80 percent of Earth's biodiversity.

The World Resources Institute found that Indigenous and similar community lands with secure tenure (i.e., rights to use the land) are linked to low deforestation rates, significant forest cover, and sustainable production of timber and other products. The combined 20 year economic benefit was US$700 billion to 1.56 trillion. This value includes US$25-34 billion for avoiding an annual release of 48 to 60 metric tonnes of carbon dioxide emissions. Bolivian, Brazilian, and Colombian Indigenous lands emit 73 percent less carbon than lands managed by others.

Indigenous Communities Have Rights

Thirteen years ago, the United Nations General Assembly adopted the Declaration on the Rights of Indigenous Peoples. Today, 148 of 193 United Nations members have adopted the Declaration, and 182 members reached consensus on an outcome document that welcomed the adoption of the Declaration. No members opposed the Declaration.

Declaration highlights include:

- Seventeen articles addressing how to protect and promote Indigenous culture by respecting its input and allowing resources for education in Indigenous languages.
- Fifteen articles about Indigenous peoples' participation in all decisions that will affect their lives.
- Confirmation that Indigenous peoples deprived of their means of subsistence and development are entitled to just and fair redress.
- Outlawing discrimination against Indigenous peoples.

- The right of Indigenous peoples to remain distinct and to pursue their own visions of economic and social development.
- Required free, prior, and informed consent of Indigenous peoples before matters of fundamental importance to their rights, survival, dignity, and well-being are decided.

The Plight of Indigenous Peoples

Despite the establishment of international Indigenous rights, Indigenous peoples continue to be marginalized at national and local levels. Many countries do not recognize Indigenous peoples within their territories. Indigenous land custodianship is often unrecognized by national laws and policies.

A significant challenge to Indigenous peoples is pressure from development, forestry, agriculture, and extraction on and adjacent to their lands. Environmental degradation impacts their livelihoods, food, water, shelter, and culture, affecting survival and health and triggering migration. Biodiversity and other ecosystem services are diminished - including carbon sequestration. Indigenous people are killed - 85 in 2019 - protecting their lands from mining, agriculture, and logging interests.

A second challenge to Indigenous rights is linked to the conservation movement. Conservation initiatives, including the establishment of protected areas, biodiversity protection strategies, forest management programs, and carbon sequestration projects, continue to confiscate Indigenous lands and evict communities. Recent violations affected the Baka in Cameroon, the Basarwa and the Mbuti in the Congo, the Maasai in Tanzania, and the Senwger in Kenya. Justice is hampered by racism, lack of political will, and the failure to recognize the rights of Indigenous peoples.

The conservation-related plight of Indigenous peoples can be traced back to the establishment of Yellowstone National Park in 1872. A concerted and at times violent effort expelled the resident Miwok Native Americans who settled the valley 4,000 years earlier. Many thousands of Native Americans were expelled to create Yellowstone, and later Grand Canyon, Mesa Verde, Mount Rainier, Zion, Glacier, Everglades, and Olympic National Parks.

The Yellowstone Model was exported around the world, contributing to around twenty million conservation refugees. The Batwa people were evicted in southwestern Uganda in the 1990s to accommodate the Bwindi Impenetrable and Mgahinga Gorilla National Parks. The evictions led to homelessness, poverty, cultural loss, and declining health. The Sengwer people were evicted in Kenya, and a person killed, to accommodate a water management project funded by transnational conservation. In Bangladesh, traditional livelihoods were discouraged, and often prohibited, in newly established protected areas.

Some Indigenous peoples, especially refugees, believe that conservation serves the needs of wealthy outsiders, and they conflate conservation with imperialism. At the heart of the issue is a fundamental lack of understanding and respect for divergent views. The Yosemite Model represents 'fortress' conservation, the view that nature must be free of people. This perception has persisted in some quarters to the present day, even among well-meaning conservationists.

By contrast, Indigenous peoples believe that nature is sacred and should be cared for, and that people are part of nature. Indigenous peoples have been sustainably managing their lands for hundreds of years, developing deep knowledge and effective practices that enable their lands, livelihoods, and culture. The concept of nature as separate is nonsensical and impractical.

How Indigenous Peoples Can Contribute to Sustainability

The big conservation goals required for a healthy planet (e.g., Half Earth, 30 by 30, Sustainable Development Goals, Climate Agreement) cannot be achieved without Indigenous peoples. Most lands warranting biodiversity conservation, climate change action, and ecosystem service maintenance are held or managed by Indigenous groups. So too, generations of Indigenous peoples have been observing the environment and recognizing patterns, the very information needed to be good stewards. Information as important as scientific studies.

Governments, nongovernmental organizations, and multilateral banks have recognized in recent years the criticality of engaging and empowering Indigenous groups. While not ubiquitous, national and local governments are beginning to recognize the rights of Indigenous groups to own and manage land and contribute to decision-making. Transnational nongovernmental organizations are collaborating with Indigenous groups. And multilateral banks are developing frameworks and policies that link environmental and social issues, including the rights of Indigenous groups.

The contributions of Indigenous groups are significant and varied illustrating the benefits of engagement and respect.

- Coastal Shellfish, a sustainable, Indigenous-owned and operated company in British Columbia, Canada is farming scallops in the natural environment and plans to add kelp and sea urchins.

- The Ogiek people, evicted from the Mau forest complex in Kenya in the 1930s, have returned to their homelands and are planting indigenous trees to restore deforested areas.
- In Nicaragua, sixty-one natural reserve areas are categorized as flexible conservation areas allowing local populations to use natural resources for their direct benefit.
- In Canada, the Indigenous Circle of Experts supported the country's efforts to protect at least 17 percent of its lands and 10 percent of its marine waters by 2020.
- In Arizona, the Havasupai Tribe administers conservation lands adjacent to Grand Canyon National Park and the Navajo Nation administers Monument Valley Tribal Park.
- In Finland, 15,000 square kilometers are designated as wilderness to protect wildlife and Sami culture, including traditional subsistence methods.
- In New Zealand, the Māori Tribes won a 140 year old quest to recognize the Whanganui River as an ancestor, achieving lawful protection from degradation.

Indigenous groups developed a respect for nature and stewardship long before the conservation movement began. They have observed environmental changes and detected patterns for generations. Indigenous peoples possess the complementary knowledge and expertise to enhance the conservation practices needed to address climate change, biodiversity loss, and the diminishment of ecosystem services.

Actions to Sustain the Planet and Preserve Traditional Knowledge

- Adopt the paradigm that conservation can be compatible with, and enhanced by, human activities, especially the traditional livelihoods of Indigenous peoples.
- Cease sole reliance on published scientific materials and demonstrate equal respect for the knowledge of Indigenous peoples.
- Secure the land rights of Indigenous peoples to advance climate change mitigation and stem biodiversity loss.
- Collaborate with Indigenous peoples to develop innovative methods for conservation.
- Include Indigenous peoples in decision-making at international, national, and local levels.
- Encourage Indigenous peoples to create their own protected areas.
- Place tenure-secure Indigenous lands at the center of conservation strategies.
- Sensitize stakeholders to potential conservation impacts to Indigenous peoples.
- Institutionalize free, prior, and informed consent for any activities with the potential to impact Indigenous peoples.

HEALTHY LANDS AND WATERS CAN REDUCE PANDEMIC RISK

Our continued erosion of wild spaces has brought us uncomfortably close to animals and plants that harbor diseases that can jump to humans.

Ingrid Andersen, Executive Director of the United Nations Environment Programme.

Josef Settele, Sandra Diaz, and Eduardo Brondizio of the Intergovernmental Science-Policy Platform on Biodiversity and Ecosystem Services, and Peter Daszak of EcoHealth Alliance, issued a statement about the interdependence of pandemics and nature. Seventy percent of all emerging human diseases originate in wildlife and domesticated animals.

Deforestation, expansive agriculture, intensive farming, mining, infrastructure development, and wild species exploitation bring people into contact with wildlife-borne diseases causing 700,000 deaths each year. Any one of 1.7 million unidentified viruses could trigger the next pandemic.

Ending the COVID-19 pandemic and avoiding future pandemics requires change. The authors emphasize that avoiding the pandemic cycle will require system-wide reorganization. But the cost benefit ratio is in our favor.

- Strengthen and enforce environmental regulations and deploy economic stimulus packages that offer incentives for sustainable activities.
- Adopt a "One Health" approach that recognizes the interdependence among human and environmental health.
- Fund and support healthy systems that incentivize change and build health capacity in emerging hotspots.
- Offer viable and sustainable alternatives to high risk economic activities and protect the most vulnerable individuals.

CORONAVIRUS, CHANGE, AND DOING BETTER NEXT TIME

Some people don't like change, but you need to embrace change if the alternative is disaster.

Elon Musk, technology entrepreneur

Elon Musk was not referring to the coronavirus when he made the above statement, but it is applicable. The coronavirus pandemic is the greatest change many people will face in their lifetime. Or so we hope. Realistically, the tension between the built and natural environments will continue for some time resulting in more wildlife to human disease transmission.

The coronavirus pandemic has much to teach. An extended or reoccurring coronavirus, or new pandemics, can be neutralized or minimized if we learn its lessons and embrace change.

Lessons

- Leadership is paramount. Integrity, competence, forward looking, humble, and service to others are the qualities required of leaders during a pandemic.
- Pandemics cannot be neutralized by wishful thinking.

- Politicians rarely have the necessary training and experience to design and implement an effective response to pandemics.
- Divisiveness and partisanship diminish the success of any pandemic response.
- The elderly, underserved, and disenfranchised are the most vulnerable to coronavirus and future pandemics.
- Close contact with wildlife facilitates the transmission of disease and the rise of pandemics.

Embracing Change

Learning these lessons requires change. Change is constant. Change is a necessary and a critical life skill for adapting to a rapidly changing world. But change is difficult for many. It challenges our propensity for habits triggering our fight or flight response. Change challenges our intellectual and emotional capacity.

During pandemics leaders must earn trust by being honest and transparent. They must normalize change and help others embrace change. They must demonstrate empathy for the plight of others.

Leaders can look to Elizabeth Kubler Ross's change curve to help them understand and recognize their own and others' emotional states. Understanding and recognizing emotional states confers some control of everyone's response to the change needed to survive a pandemic. The curve also reminds us that we are not alone - change impacts everyone. And finally, the curve teaches us that our shock and denial, anger, bargaining, and depression will eventually lead to acceptance and integration of the change into our lives.

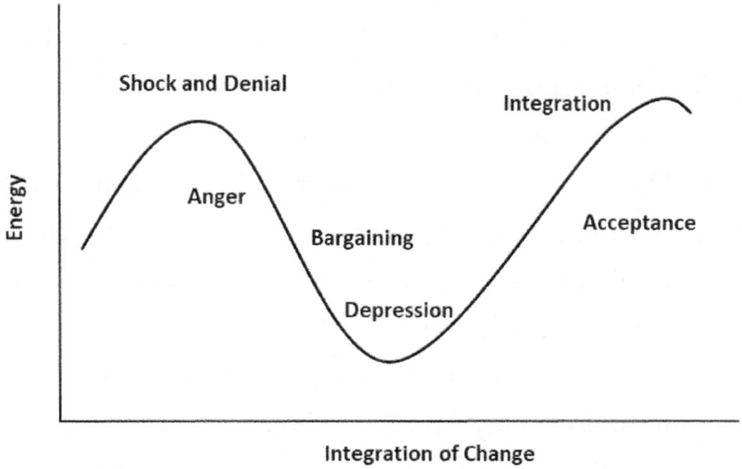

Source: Adapted from Elisabeth Kubler-Ross and David Kessler. 2014.

Doing Better Next Time

With lessons learned and an understanding of change we can do better next time. Here are some actions we can take to ensure the next pandemic is neutralized or minimized.

- Elected and appointed officials should be evaluated for their experience responding to crises, capacity for leading others through change, and honesty. Their ability to recognize their limitations and to rely on experts to guide decision-making and communication with the public. Our leaders should be humbled by the responsibility that comes with service to others, and avoid blaming others or practice partisan politics, but stay focused on the immediate task.

- Hold on to the moral compass. Talk of sacrificing some people to benefit others increases fear and undermines our ability normalize change and neutralize a pandemic's impact. Every life is equally valuable.
- Develop a national strategic plan for the next pandemic. Determine who should be part of the team, the team's structure, and the decision-making process. Develop scenarios based on potential types of pandemics, where the pandemic might begin, time needed to respond, and needed facilities, medical equipment, laboratory reagents, and trained health care providers. Identify mechanisms for protecting health care workers and other essential personnel.
- Integrate the national strategic plan with those of other countries and global health organizations. Global pandemics require collaboration, coordination, and good will. Collective intelligence and optimal capacity are paramount. In a pandemic, there is no here and there or us and them.
- The elderly, under-served, and disenfranchised are the most vulnerable to the coronavirus pandemic and other pandemics. They often lack health insurance, savings, and are asked to provide essential services. Develop a plan to ensure the elderly, under-served, and disenfranchised have medical care, food, and shelter, and funds to pay rents and mortgages if they are out of work. Provide enhanced protection for those who continue working in essential facilities like take- out restaurants and grocery stores.
- The best way to neutralize a pandemic is to stop it from starting. Many pandemics arise from close human contact with wildlife. Contact occurs when people rely on wildlife for food,

encroach on wildlife habitat for agriculture, or share unsanitary conditions with wildlife. We must develop social-ecological systems that ensure that all people are able to avoid close contact with wildlife. And we must ensure our natural environments are healthy to minimize the development of disease in wildlife.

- Practice patience, tolerance, and compassion. People suffer in pandemics. Be patient with those who struggle. Tolerate circumstances you normally would not. And show compassion for those less able to cope in a pandemic. Ask how you can help not only yourself, but others.

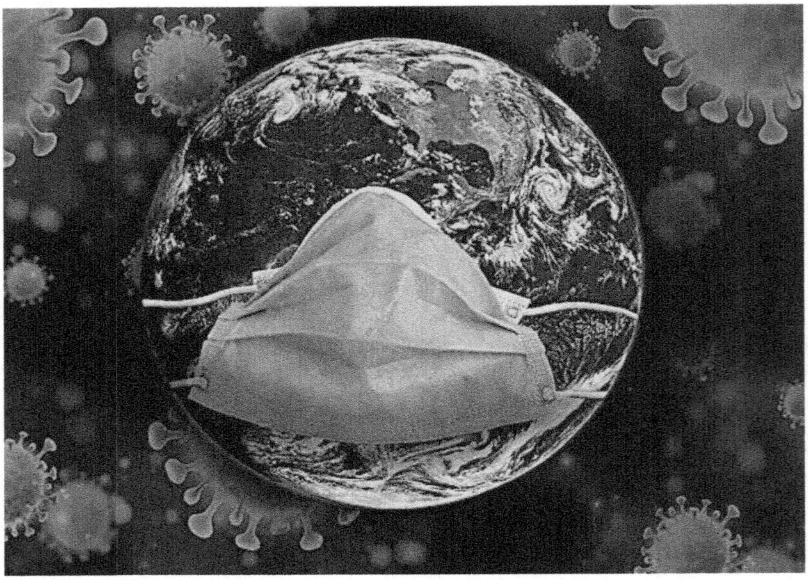

ENSURING FOOD SECURITY DURING PANDEMICS

The COVID-19 pandemic is a health and human crisis threatening the food security and nutrition of millions of people around the world. Hundreds of millions of people were already suffering from hunger and malnutrition before the virus hit and, unless immediate action is taken, we could see a global food emergency.

United Nations Policy Brief: The impact of COVID-19 on food security and nutrition.

The recent COVID-19 pandemic exposed the fragility of our global food systems. About 768 million people faced hunger in 2020 during the pandemic, 118 million more than in 2019. Projections indicate that without change 660 million will face hunger in 2030 - 30 million more than if the pandemic had not occurred.

Pandemic-related border restrictions and lockdowns slowed harvests, eliminated the livelihoods of seasonal workers, and constrained food transport. Unable to get product to market, farmers buried perishable produce and dumped milk.

Hundreds of millions of school children missed daily meals at school, meals critical to their health. A June 2020 report from the

Global Network Against Food Crises found that at least 135 million people in fifty-five countries were suffering from a food crisis before the pandemic. Experts predicted the number would double by the end of the year due to COVID-19.

High levels of unemployment, loss of income, and rising food costs reduced access to food for people of lesser means. Tens to hundreds of millions were expected to experience extreme poverty by the pandemic before the year's end. Diminished cash flow and liquidity would force individual producers and small agribusinesses to cease operations.

Minimizing Impacts on Food Supplies During Pandemics

Concerted action is needed to minimize the food security impacts and to enhance food system resilience to future pandemics. Here is a synopsis of the United Nations framework for enhancing food security during the next pandemic.

Mobilize collective resources to save lives and livelihoods focusing where risk is most acute.

- Continue humanitarian food, livelihood, and nutrition assistance to vulnerable groups.
- Declare food production, marketing, and distribution essential services, ensure the protection of the workers, and keep trade corridors open.
- Expand food security monitoring systems to measure pandemic effects and guide priority actions.

Ensure relief and stimulus packages reach the most vulnerable.

- Strengthen social protection systems.

- Keep food and nutrition assistance at the heart of social protection programs.
- Strengthen health system response for nutritional care.
- Protect the most vulnerable groups, including women playing key roles in household and essential services delivery.
- Design nutrition-sensitive social protection programs to ensure access to diverse, balanced, and nutritious meals.

Invest in a sustainable future.

- Transform food systems so they work better with nature and the climate.
- Lay the foundation for a more inclusive, green, and resilient recovery.
- Use the 2021 Secretary-General-hosted food summit for inclusive dialogues and mobilizing multi-stakeholder actions.

Reducing the Incidence of Pandemics

The effective way to mitigate the impact of pandemics on food supplies is to avoid pandemics. Two new viruses per year have infected people over the last century. The infections occur when people or their domestic animals come in to contact with wildlife. The potential for virus transfer is a function of contact rate, the abundance of susceptible humans or livestock, and the abundance of infected wild hosts. Much of the contact is attributed to hunting, the trade of wildlife for food, pets, and medicine, and deforestation for forest products, agriculture, mining, and development.

Three actions can significantly reduce pandemics and avoid disruptions to the food supply.

- Prevent or make safe the trade of wildlife species that have a high risk of spreading disease, beginning with bats and pangolins. Many of the virus outbreaks can be tracked to markets that sell wild animals.
- Reduce deforestation that increases contact between people, their domestic animals, and disease-carrying wildlife. Human encroachment into undisturbed forests exposes people to wild animals and their diseases.
- Invest in research programs to detect disease origination and spread. Human exposure to disease is currently underreported.

The combination of the United Nations framework and reducing pandemics through eliminating the wildlife trade, maintaining forests, and early detection can effectively reduce the number and severity of pandemics and their impact on food supplies.

COVID-19 FORCES ISLANDS TO RETHINK FOOD SECURITY

I must assure [the local population that] food is readily available and affordable.

Mayor Michael Lim Orayani, Lubang, Philippines

The pandemic was difficult for most, but for islands it was a startling reminder of their dependence on others for food. Islands import food because they lack arable land, are challenged by pests and disease, rely on more lucrative economic sectors, or have long histories of importing food. A pandemic-forced lockdown and interruptions in deliveries is driving a re-examination of agricultural policies.

Cook Islands

The Cook Islands, heavily dependent on New Zealand for food, is shifting to agro-ecological farming. A pandemic recovery package provided vegetable seedlings to 394 households, offered a free tractor service, and encouraged planting low maintenance fruit trees.

Lubang, Philippines

In Lubang, Philippines, travel restrictions isolated the island from the mainland. The government formed a partnership with local fishers asking them to catch more fish for island residents.

Puerto Rico

In Puerto Rico, which receives 85 percent of its food from the US, the nonprofit World Central Kitchen played an important role. The organization provided daily meals for essential workers and community groups, and collaborates with local farmers, fishers, and agribusinesses to ensure food remains local. World Central Kitchen also helped small holding farmers and agribusinesses purchase infrastructure and supplies, collaborate, and transition to sustainable business models.

EUROPEAN FOOD SECURITY POLICY

Food security is an authentically human requirement. Guaranteeing it for present and future generations also means safeguarding ourselves against the uncontrolled exploitation of natural resources. Indeed, the process of consumption and waste seems to overlook any concern for ... biodiversity, which is so important for agriculture.

Pope Benedict XVI

The European Union (EU) launched the Common Agricultural Policy (CAP) in 1962 as a partnership between agriculture and society and Europe and its farmers. CAP recognized that agriculture is an essential element of society but vulnerable to low wages, weather and climate, and lags between consumer demand and supply. CAP's premise is that farmers should be able to work in a sustainable and environmentally friendly manner if we are to sustain soils and biodiversity.

CAP's goals include:

- Supporting farmers and improving agricultural productivity to ensure a stable supply of affordable food.
- Safeguarding European Union farmer's ability to make a reasonable living.

- Helping tackle climate change and the sustainable management of natural resources.
- Maintaining rural areas and landscapes.
- Keeping the rural economy alive by promoting jobs in farming, agri-food industries, and associated sectors.

CAP is managed and funded in the EU budget via three categories.

- *Income support* through direct payments ensures income stability, rewards environmentally friendly farming, and the delivery of public goods not paid for by markets.
- *Market measures* address reductions in demand due to unforeseen circumstances like pandemics or temporary oversupply on the market.
- *Rural development measures* address specific needs and challenges facing rural areas.

To assess CAP's progress, the European Commission developed a common monitoring and evaluation framework. The Commission also regularly consults civil groups, agricultural committees, and experts, and conducts impact assessments.

Farm to Fork

The complementary EU Farm to Fork Strategy is a critical part of the European Green Deal designed to make Europe the first climate-neutral continent by 2050. Farm to Fork addresses the challenges of a sustainable food system and the links between healthy people, healthy societies, and a healthy planet. An updated CAP facilitates Farm to

Fork by helping farmers to improve their environmental and climate performance.

The EU's Farm to Fork goals include:

- Reduce the environmental footprint of the EU food system.
- Strengthen the food system's resilience.
- Ensure food security in response to climate change and biodiversity loss.
- Lead a global transition towards competitive sustainability by tapping into new opportunities.

Achieving these goals requires ensuring the food chain has a neutral or positive environmental impact, and everyone has access to sufficient, nutritious, affordable, and sustainable food. All actors in the food chain need to cooperate and coordinate. In addition, the EU recognizes that food system sustainability is a global issue, and it intends to play a key role in setting global standards.

Farmers are expected to use bio-fertilizers and renewable energy, reduce reliance on chemical pesticides through integrated pest management, and eliminate excess nutrients through integrated nutrient management. Also, farmers are expected to reduce greenhouse gas emissions from livestock operations, reduce antimicrobial use, and improve animal welfare. The EU will reinforce vigilance on plant imports to prevent the introduction of pests and diseases and to secure and maintain seed diversity.

Other anticipated actions include:

- Coordinating a common European response to crises affecting food systems.
- Developing a code of conduct for responsible business and marketing practices.
- Promoting healthy, sustainable diets.
- Reducing food waste.

Common Fisheries Policy

The EU's related Common Fisheries Policy will increase efforts to bring fish stocks to sustainable levels, and adopt guidelines for sustainable aquaculture development. The EU will work with Mediterranean countries to strengthen fisheries management and re-assess climate change risks.

Europe's transition to a sustainable food system will be supported by research and development, investment, knowledge sharing, and the pursuit of green alliances.

FOOD SECURITY REQUIRES SUSTAINABLE AGRICULTURE

A sustainable agriculture is one which depletes neither the people nor the land.

Wendell Berry, novelist, poet, essayist, environmental activist, cultural critic, and farmer

Food security - the availability of, access to, capacity to use, and stability of food - is a critical and fundamental challenge to the well-being and prosperity of people throughout the world. A third of the world's soil is severely degraded due to unsustainable agricultural practices. Agricultural desertification and land degradation cost about $490 billion per year. More than eight hundred million people do not have enough food to eat, and nine million people die of hunger and hunger-related diseases every year.

At its best, agriculture and mariculture are allies in the fight against climate change and biodiversity loss. Each can function as a sink or a source for greenhouse gasses, and a conservator or reducer of biodiversity. At its worst, agriculture is responsible for about a fourth of global greenhouse gas emissions, 70 percent of freshwater use, and 80 percent of habitat loss. Pesticides and deforestation for agriculture are major contributors to biodiversity loss on land and in sea.

Food security drives unsustainable agriculture, which in turn drives climate change and biodiversity loss. Climate change and biodiversity loss can only be reversed if we address hunger. We operate in a social-ecological-economic-political system. Solving one problem requires solving all problems.

What is the Future of Farming

The International Union for Conservation of Nature (IUCN) released *Approaches to Sustainable Agriculture: Exploring the Pathways Towards the Future of Farming* in 2020. The purpose of the report is to demystify and clarify some of the many terms relating to sustainable agriculture.

IUCN identified fourteen approaches with clear principles and environmental, economic, and social objectives.

1. *Agroecology* considers agriculture and the ecosystem as one system with physical limits. Nutrients and energy in the system are recycled and optimized. Biological interactions and synergies are enhanced to promote ecological processes.
2. *Nature-inclusive agriculture* is based on agroecological principles, but seeks to become mainstream in contrast to the local food chain context of agroecology. Nature-inclusive agriculture minimizes the negative effects of agriculture on nature and maximizes the positive effects of nature on agriculture.
3. *Permaculture* designs and maintains agriculturally productive ecosystems with the diversity and resilience of natural ecosystems. The goal is the integration of landscape and people, providing food, energy, shelter, and other needs in a sustainable way.

4. *Biodynamic agriculture* views the farm as a self-contained and self-sustaining organism. Farmers avoid synthetic pesticides and fertilizers and set aside a minimum of 10 percent of the farm for biodiversity.
5. *Organic farming* is a production system that sustains soil, ecosystem, and human health by relying on ecological processes, biodiversity, and cycles adapted to local conditions.
6. *Conservation agriculture* conserves the soil, enhancing biodiversity within and on the soil, capturing carbon, and enabling water and nutrient efficiency. Food production is improved and sustained.
7. *Regenerative agriculture* enhances and sustains soil health by restoring organic matter, thereby boosting fertility and productivity. Agricultural yields are increased and are more resilient to climate change.
8. *Carbon farming* is based on the principles of conservation and regenerative agriculture. The approach implements practices that improve the rate carbon dioxide is removed from the atmosphere and converted to plant material or soil organic matter.
9. *Climate-smart agriculture* increases the capacity of agricultural systems to support food security by incorporating adaptation to climate change. The approach seeks to increase agricultural productivity and incomes, enhance resilience and adaptation to climate change, and reduce greenhouse gas emissions.
10. *High nature value farming* recognizes that biodiversity is dependent on, and intertwined with, traditional, low-intensity farming. The approach encompasses traditional farming

practices like pastoralism (extensive grazing) and low-input, small-scale cropping.

11. *Low external input agriculture* strives to reduce the use of inputs from outside the production system and optimizing the management and use of internal production inputs. The net effect is a balanced and efficient use of nutrients, organic matter, water, energy, and genetic resources.
12. *Circular agriculture* emphasizes the use of arable land to produce plant biomass for human consumption. External, finite resources are minimized encouraging the use of regenerative resources. Carbon, nitrogen, phosphorous, and other resources are reused and recycled.
13. *Ecological intensification* satisfies the anticipated increase in food demand while meeting environmental quality standards. The approach relies on biodiversity conservation, improved soil fertility management, better ecological understanding to reduce pests and disease, and resilience through plant diversity and adaptation,
14. *Sustainable intensification* seeks to increase yields without adversely impacting the environment or cultivating more land. Practices draw from agroecology, organic farming, precision farming (i.e., robotics, artificial intelligence, and big data), urban farming, genetic improvement, and greenhouses.

The approaches vary in their emphasis on yield and relationship to nature, but all are considered sustainable and useful in the right circumstances. IUCN defines each approach and describes its principles, practices, challenges, and opportunities. Many of the approaches share practices, - crop rotation, cover and companion crops, mixed crop and inter-cropping, minimal tillage, reduced pesticide and fertilizer use, and lower livestock densities.

Acceptance, scalability, and profitability challenge a transformation from conventional to sustainable agriculture. A common vision, the development of joint metrics, and governmental support for transforming the global economy are required for success.

There is no single definition of sustainable agriculture, but most agree that the needs of both current and future generations must be met and that actions consider environmental, social, and economic sustainability. Regardless of the precise definition, sustainable agriculture must encompass not only in-place practices but the cost of processing and transport.

WHY WHAT WE EAT MATTERS

A global shift in diets to prioritize nutrition hand-in-hand with sustainability - something that will look different in every country - has the potential to reverse biodiversity loss, combat climate change, and save human lives.

Brent Loken, World Wildlife Fund global food lead

Food Vulnerabilities

The combination of increasing productivity and the absence of a major natural or anthropogenic event have produced enough food to feed the planet's 7.8 billion residents. If all goes well, the food supply will increase by 20 percent over the next decade, producing more than enough food to supply the burgeoning global population.

This rosy picture obscures several significant vulnerabilities.

- Food is not equitably distributed, and middle income and wealthy people consume more calories, meat, and other animal-source food than is required for nutrition.
- Our agricultural production and diets lack diversity. We are largely dependent on only ten crops despite 6,000 crops being historically used for food. Homogeneity hampers our

adaptability and increases food susceptibility to pests, disease, natural hazards, and climate change.

- Global food production is geographically concentrated. Sixty percent of food production occurs in concentrated regions of China, the United States, India, Brazil, and Argentina. The likelihood of a production center failure occurring in any given year has increased from 1 percent in the past 20 years, to 2 percent in the next decade, and 4 percent by 2050. An impact to production could raise prices by 100 percent, impacting 750 million people living below the international poverty line. Social unrest could ensue.
- Some countries, including many developing countries, are dependent on imports that rely on functioning, organized trade relationships and geopolitical stability. Shortages are already occurring due to interruptions in agricultural labor, export bans, trade wars, natural hazards, political conflict, and pandemics. Vulnerable developing countries tend to be importers of essential grains.
- The economic maturation of developing countries is shifting diets toward higher meat consumption. Increased demand threatens food availability and affordability while straining ecosystems.
- Our ability to feed the global population is only achieved by exceeding planetary boundaries - the safe operating space for humanity relative to Earth's system processes. Adhering to the boundaries will only feed 3.4 billion people. Progressive climate change will exacerbate the situation, pushing food crops beyond their physiological limits.

Threats to Human Health

Food-related illnesses are the leading cause of death. About nine million people die of hunger and hunger-related diseases every year.

No country is on course to meet 2025 global nutrition goals. More than 850 million people, one in every nine, are chronically undernourished. Millions of children are stunted and wasting due to poor nutrition.

Most of the world's population lives in countries where cardiovascular disease, hypertension, diabetes, stroke, and cancers linked to overconsumption kill more people than hunger. Two billion people are overweight or obese. The combination of increased food intake and decreased activity has tripled obesity since 1975. Unhealthy diets pose a greater risk of disease and mortality than unsafe sex and alcohol, drug, and tobacco use combined.

Threats to Ecosystem Health

Land-use change, primarily conversion for agriculture, is the most important driver of biodiversity loss.

- Agriculture occupies 50 percent of habitable land and is responsible for 80 percent of habitat loss and 70 percent of freshwater withdrawals.
- Food systems are the largest polluter of water resources. Agricultural land conversion and pollution are responsible for a 70 percent decrease in terrestrial biodiversity and a 50 percent decrease in aquatic biodiversity.

- Agriculture accounts for 80 percent of global deforestation. Conversion for beef livestock, palm oil, and soy account for 70 percent of tropical deforestation
- Agriculture is a principal driver of climate change. Twenty-seven percent of greenhouse gases emanate from agricultural activities. Emissions are expected to double this century.
- Wildlife is consumed by more than half the world's population. Overexploitation of wildlife is exacerbating biodiversity loss.
- Only 66 percent of fish stocks are sustainable, a 27 percent decrease in sustainability between 1990 and 2017.
- Freshwater fisheries are under increasing threat from overfishing, pollution, and unsustainable development.
- Aquacultural expansion is associated with inland conversion, mangrove forest destruction, parasite and disease spread, the introduction of nonnative species, the misuse of chemicals, and pollution.

Improving Human and Ecosystem Health

Supply Side Solutions

Sustainable and regenerative agriculture are supply side solutions to improving human and ecosystem health. Supply side solutions encourage sustainability, land restoration, and environmental sensitivity while preserving livelihoods. Supply side change is important, especially for sustaining the environment. But without aggressive regulation, supply side solutions alone are inadequate for promoting the necessary positive feedback to improve human and ecosystem health.

A Demand Side Solution

The demand side of the equation - modifying consumer behavior - is a complementary opportunity to transform food systems and improve human health and ecosystems. The opportunity encompasses increasing consumer acceptance and demand for plant-based foods, especially protein-rich foods, and decreasing demand for animal-based foods. Plant-based foods tend to have a lower planetary impact than animal-based foods. Animal-based foods do not need to be eliminated, only reduced, to achieve gains in sustainability and satisfy nutritional, cultural, and ecological needs.

Change

People have and will change their diet. Forty years ago, each person in the United States ate ninety pounds of red meat and forty pounds of chicken. Now, each person eats sixty pounds of red meat and ninety pounds of chicken.

In the United States and United Kingdom, people report being motivated to adopt plant-based food products by concerns for their health, the environment, and animal welfare. They are substituting plant-based products for fast- and processed foods. Asians adopted plant-based proteins centuries ago.

Ron Salpeter, CEO of superfood-maker Mankai, a complete protein derived from an aquatic plant, sees two dietary trends shaped by the demands of millennials and Gen Zers. Conscious Eating considers how food affects the body and the planet. The Plant Forward Agenda ensures people are exposed to fruits and vegetables because of elements lacking in processed food.

Plant-based Proteins

Soy-based (e.g., tofu) and to a lesser extent, pea-based proteins have been universally accepted by the public. According to Innova, new food offerings with vegan-related claims have grown more than 45 percent in the last few years. Impossible Foods gained notoriety for two years for its impossible burger. Several other plant-based meat alternative companies are up and running with at least one successful Initial Public Offering. Kellogg, Purdue, Tysons, Nestlé, and Smithfield Foods are established meat companies deeply invested in plant-based alternatives.

Plant-based protein alternatives are more sustainable and far less expensive than meat. Soy, pea, whey, and fungal proteins cost between US$0.02-0.13 per kilogram to produce. By comparison, chicken is about US$2.00 per kilogram and grass fed beef US$3.20 per kilogram. Cultured meat, produced by in vitro cell culture of animal cells, costs US$300 per kilogram.

National Dietary Guidelines and Labelling

The development and application of universal and national dietary guidelines is a necessary adjunct to behavioral change, as many people will accept expert knowledge and medical advice. Adoption of guidelines emphasizing plant-based foods will be aided by cost advantages. Also, the guidelines will help eliminate the stigma of plant-based diets being associated only with low income individuals or food progressives. Food labelling that indicates bioavailability (i.e., how much is absorbed by the body) and impact on the environment will appeal to younger consumers.

Improved Sustainability and Human and Ecosystem Health

What we eat does matter and can improve food system sustainability and human and ecosystem health. The key is greater reliance on plant-based foods, including proteins.

- Agricultural land requirements and costs can be reduced.
- Human health will be improved by increasing the amount, availability, and nutritional value of food, reducing starvation, malnutrition, illness, and disease.
- Ecosystem health will be improved by converting land historically used to produce animal-based food to cropland and restoring excess land to a natural state.
- Land restored or that avoided conversion to agriculture would store carbon, conserve biodiversity, recycle nutrients and wastes, and mitigate water pollution.

Sustainable, regenerative agriculture coupled with dietary change can transform our food systems. Reducing the costs of production and consumption, enhancing adaptability, improving human health, and returning lands to nature.

Incidentally, the ten crops we rely on are corn, potatoes, sweet potatoes, yams, cassava, soybeans, sorghum, plantain, wheat, and rice.

HOW TO REGENERATE ECOSYSTEMS WHILE GROWING FOOD

The health of soil, plant, animal, and man is one and indivisible.

Sir Albert Howard, English botanist and organic farming pioneer

Regenerative Agriculture

The modern concept of regenerative agriculture was articulated at the Rodale Institute more than 50 years ago. Soil health was the focus, although Robert Rodale intended the concept to encompass a broad spectrum of human values.

Regenerative agriculture aims to improve land by:

- Contributing to soil building, fertility, and health.
- Increasing water percolation, retention, and clean, safe water runoff.
- Enhancing and conserving biodiversity.
- Building ecosystem capacity for self-renewal and resiliency.
- Sequestering carbon.

Traditional practitioners minimize or avoid tillage, use cover crops, apply compost, compost abstracts, and animal manures, inter-crop diverse plants, and establish borders for bees and other beneficial insects.

Regenerative Grazing

Regenerative grazing is a principle-driven, flexible, and adaptive agricultural practice that manages livestock on perennial and annual forages to support human and ecosystem health, farm profitability, and community and food system resilience.

Regenerative grazing practices dispense with feed lots and confined animal feeding systems, stimulate plant growth, increase soil carbon sequestration, and provide a healthy environment for animals. Management factors include the shape and size of the grazing area, stock density, length of time an area is grazed, how often the area is grazed, and time of year.

Regenerative Agroforestry

This ancient practice dates back at least 4,500 years to the Spanish Denesas system of cattle, pig, and sheep pasture mixed with scattered oaks. Regenerative agroforestry was also represented in Germany until the Middle Ages, in the pre-Columbian Americas, the Indian peninsula, and Africa.

Regenerative agroforestry mimics nature, integrating trees with complementary crops or animals. Compared to monoculture, regenerative forestry increases resilience, biodiversity, productivity, and profitability while lessening climate change impacts. Agroforestry's core principles include:

- Covering soil with biomass and organic matter.
- Diversified plantings.
- Growing food crops.
- Exchanging knowledge among practitioners.

- Economic feasibility.

Agroforestry systems can be agrisilvicultural (crops and trees), silvopastoral (trees and animals), or agrosilvipastoral (crops, trees, and animals).

The Nature Conservancy's Regenerative Agriculture Efforts

The Nature Conservancy (TNC) has an expansive approach to regenerative food systems. Their goal is to produce food on land and sea using practices and policies that restore habitat, protect biodiversity, and reduce greenhouse gas emissions while preserving the livelihoods of farmers, ranchers, and fishers.

Achieving that goal requires changing production, processing, distribution, and retail practices through science, corporate action, and policy. Two TNC exemplars are their partnership with Syngenta, and their shellfish and seaweed aquaculture projects.

Syngenta and The Nature Conservancy

Syngenta launched its new *Good Growth Plan* in June of 2020 to reduce agriculture's carbon footprint and to help farmers' build resilience to climate change-related extreme weather. The Plan includes a commitment to invest $2 billion in sustainable agriculture by 2025, and deliver two technological breakthroughs each year.

Syngenta's goals are to accelerate innovation for farmers and nature, strive for carbon neutral agriculture, and help people be safe and healthy. The Plan is a continuation of the first *Good Growth Plan* in 2013 which recovered fourteen million hectares of farmland and enhanced biodiversity on more than eight million hectares.

Regenerative Aquaculture

The Nature Conservancy's regenerative aquaculture efforts extend around the world, offering the potential to increase food production while facilitating coastal ecosystem recovery. Bivalve shellfish and seaweed aquaculture improve water quality by assimilating nutrients and support improved food security, employment, and gender equality. Bivalve shellfish provide fish habitat and enhance benthic community diversity and productivity. Seaweed aquaculture reduces the local effects of ocean acidification by taking up carbon dioxide.

A recent study led by The Nature Conservancy determined that Oceania (Australasia, Melanesia, Micronesia, and Polynesia), North America, and parts of Asia have the greatest opportunity for shellfish aquaculture. Europe, Asia, Oceania, and North and South America have the greatest opportunity for seaweed aquaculture.

ADAPTING FOOD SYSTEMS TO CLIMATE CHANGE

We have to change how we produce and consume food, not just for environmental reasons, but because this is an existential issue for humans.

Janet Ranganathan, vice-president for science and research at the World Resources Institute

Climate Change Will Exacerbate an Already Ineffective Food System

Our food systems are not wholly effective despite marked improvements in productivity over the last several decades. More than eight hundred million people are undernourished. More than six hundred million women and girls 15 to 49 are iron deficient, and more than 150 million children under five are stunted.

Climate change-related temperature increases, shifts in precipitation patterns, and increasingly frequent extreme climate events exacerbate food system ineffectiveness. Conversely, our food systems exacerbate climate change by contributing 21 to 37 percent of total greenhouse gas emissions.

Shifting and Declining Agriculture

Climate change is already affecting food security. Yields of maize and wheat crops are declining in lower latitude regions and increasing in higher latitude regions. Warming and drying are decreasing yields in parts of the Mediterranean, drier areas of Africa, and high mountain regions of Asia and South America.

Continued climate change is predicted to increase the cost and nutritional quality of essential cereal crops, diminish food access and availability to low-income consumers, and increase world hunger. Pastoral systems - ranching, livestock farming, and grazing - will suffer lower pasture and animal productivity, reproductive dysfunction, and biodiversity loss. Fruit and vegetable production will decline due to higher temperatures in tropical and semi-tropical regions.

Shifting and Declining Marine Ocean Fisheries

Acidification

Ocean acidification will alter marine food webs. Oceans absorb atmospheric carbon dioxide, serving as a buffer to climate change but at the cost of increasing acidity. Ocean acidification will dissolve shrimp, oyster, coral, and zooplankton exoskeletons. Changes in the abundance of these species will alter the entire marine food web.

Warming Waters

Warmer ocean waters will alter the distribution, productivity, and species composition of fisheries. Increasing water temperatures cause fish to move toward the poles or to deeper waters in search of cooler waters. Warming waters in the northeast United States have already driven fish further north over the last 20 years. Ecological shifts here and elsewhere disrupt predator-prey dynamics and coastal community economies.

In the Pacific and Caribbean, coral reefs are bleaching and dying from warming seas. Coral reefs are home to 25 percent of all marine species. About forty million people rely on coral reefs for seafood.

Sea Level Rise

Coastal marshes and mangrove forests are nurseries to countless shrimp, oysters, crabs, and fish species. Nurseries are inundated and productivity diminished by sea level rise. Every hour Louisiana, United States loses 57,600 ft^2 (5,351 m^2) of coastal wetland nursery areas significantly impacting food supply and coastal community economies.

Countering Climate Change Impacts to Food Security

Adapt Agriculture and Fisheries

Adapting agricultural systems is required to counter climate change. Cropping systems can be adapted to sequester soil carbon, and reduce nitrous oxide and methane emissions. Livestock system ratios of milk and meat to methane emissions can be improved by using higher quality feed. Better pasture management practices can increase primary productivity and soil carbon sequestration while reducing manure emissions.

Fishers will need to adapt, pursuing fish stocks towards the north and south poles, fishing in deeper water, or catching species that relocated from warmer areas. Shifting to other foods and livelihoods are likely in some instances.

Reduce Food Waste

Food loss and waste, now about 25-30 percent of total food produced, must be reduced. More than one billion tons of food never leave farms, get lost or spoiled during distribution, or are thrown away in businesses and homes. Reducing food waste will negate malnourishment, hunger, and hunger-related diseases for hundreds of millions of people. Food waste is also responsible for 6 to 8 percent of greenhouse gas emissions.

Food waste can be reduced through improved harvesting techniques, on-farm storage, packaging, and transportation. Consumers can switch to healthier and more sustainable local diets.

Adopt a Multi-pronged Strategy to Enhance Food Security

Enhancing food security, while improving Earth's environmental condition, requires:

1. Improving and hastening efforts to reduce greenhouse gas emissions.
2. Removing carbon dioxide from the atmosphere.
3. Adapting local, regional, and global agriculture and marine food systems.

BUILDING URBAN RESILIENCE TO CLIMATE CHANGE

We believe in power of process: finding the harmony between land and people through design.

Kotchakorn Voraakhom, landscape architect and United Nations 2020 Climate Action Award Winner

The United Nations Global Climate Action Awards recognize innovative and transformative solutions that address both climate change and wider economic, social, and environmental challenges. The award-winning projects are some of the most practical, scalable, and replicable examples of what people, businesses, governments, and industries are doing to tackle climate change.

Kotchakorn Voraakhom, a Bangkok-based landscape architect, is a United Nations 2020 Climate Action Award Winner. Ms. Voraakhom is creating innovative, nature-based landscape solutions to make Bangkok more resilient to climate change.

Ms. Voraakhom is the founder of the for-profit LANDPROCESS landscape architecture and urban design firm and the nonprofit Porous City Network. In combination, the organizations are advancing urban resilience and adaptability to climate change while promoting the fair treatment and meaningful involvement of all people in the development, implementation, and enforcement of environmental laws, regulations, and policies.

Bangkok is an appropriate crucible for climate change action and environmental justice. Located in the Phraya River delta, fifty years ago Bangkok was largely agricultural land able to absorb seasonal flooding and monsoon rain. Today, Bangkok is a city of more than eight million people, and the surface is almost entirely impervious.

Ambient temperatures have increased in Bangkok. Flooding and stormwater pollution are commonplace, and land subsidence is exacerbating the threat of inundation from sea level rise. Less prosperous residents occupying low-lying areas are disproportionately affected.

LANDPROCESS

Established in 2011, LANDPROCESS defends and expands carbon-sequestering landscapes to enable community resilience and adaptation to climate change. Each project is committed to productive public use and design through engagement. Examples include green roofs, water-efficient designs, urban farming and forests, wetlands, and grasslands. Two projects highlight Ms. Voraakhom's vision and achievements.

Chulalongkorn Centenary Park

Chulalongkorn Centenary Park is designed to address climate change uncertainties. The park is sloped to create a container for water and a raised green roof directs runoff from the park and neighboring areas through sloped rain gardens. Native vegetation planted in a depression temporarily holds the runoff and filters out particles. The remaining runoff passes to a constructed wetland for additional filtration, and then to a retention pond. The center of the park serves as a detention area for flooding and retention pond overflow.

Social elements of the design include herb gardens, a meditation walk, and a reading area. Two lanes of a four-lane roadway were converted to pedestrian walkways, bike lanes, and linear rain gardens that connect the park to adjoining neighborhoods.

Thammasat University Rooftop Farm

A combination of modern landscape design and traditional rice agriculture, the Thammasat University Rooftop Farm addresses water management and food security challenges caused by a burgeoning population and urban development. Rice terraces are integrated with green roof technology to produce a 237,000 ft^2 cascading rooftop, reducing runoff twenty times more efficiently than a conventional roof.

Rainwater zigzags down the roof's slope watering planted cells, purifying the water, and growing food for the University's students and employees. Four retention ponds can store three million gallons of water, reducing floods and providing water during droughts.

Porous City Network

Ms. Voraakhom established Porous City Network in 2016 with the goal of increasing climate change resilience and adaptability in vulnerable communities. Her focus is on increasing the amount of permeable landscape via a network of public green spaces. The spaces also provide social and food security services.

Porous City Network's landscape architects and urban and environmental designers identify locations vulnerable to climate change. They then collaborate with communities to raise climate change awareness and design solutions to water management, rising temperature, and food security issues. Workshops teach the use and maintenance of co-designed public spaces. Projects include:

- Green roofs
- Retention ponds and lawns
- Green linkages
- Rain gardens
- Detention ponds
- Canal restorations
- Pocket parks
- Urban farming

Ms. Voraakhom re-visioned urban landscapes as places designed for people by people. Her designs meet needs ranging from green space to escape the built environment, food, and flood reduction to lessen the impacts of climate change. If a nature-based, people-driven approach can work in a sweltering city of eight million people, it can work anywhere.

DESTINATIONS PUSH BACK ON CRUISE SHIPS

The journey, not the arrival, matters.

T.S. Elliot, poet, essayist, publisher, playwright,
literary critic, and editor.

Large cruise ships afford passengers an affordable, less stressful way to travel to exotic locations, COVID-19, norovirus, and *E. coli* outbreaks aside. For destinations, cruise ships have been a reliable source of income. But the abundance and increasing size of cruise ships have introduced challenges causing destinations to reconsider their welcome.

Cruise Ship Impact and Tension

Spain

Barcelona had the most-polluted port in Europe with over thirty-two tons of sulfur oxide emitted in 2017. The Mayor is trying to limit the number of cruise ships to reduce water and air pollution and reduce crowding.

Mallorca was the second-most polluted port in Europe in 2017 - cruise ships emitted twenty-eight tons of sulfur oxides. Eleven thousand residents signed a petition calling for a limit of one cruise ship per day, citing pollution, overcrowding, and waste build-up.

Australia

In 2018, a cruise ship spilled more than 7,000 gallons of food waste onto the Great Barrier Reef.

Greece

Santorini's electricity grid and water supply are stressed, and garbage has doubled in the last five years. The 29^2 mile island is reducing the number of cruise ship passengers per day from 10,000 to 8,000.

Italy

The six hundred cruise ships that entered Venice every year endangered the cultural and social integrity of the city, damaged infrastructure, and caused accidents. Venice banned cruise ships weighing more than 96,000 tons from entering the Giudecca Canal that borders the city center.

Croatia

In Dubrovnik, a UNESCO World Heritage Site, visitor traffic exceeded the capacity of the city's thoroughfares. The city reduced the number of cruise ships per day from ten to two, and the number of visitors per day from more than 10,000 to 5,000.

Ecuador

In 2007, the United Nations added the Galápagos Islands to the UNESCO list of *World Heritage in Danger* over concerns cruise ships would introduce invasive species. Today, every cruise ship that enters the Galapagos Islands must be approved by the Galápagos National Park, the Maritime Authorities, and the Ecuadorian Ministry of Tourism to ensure that it is not endangering wildlife and the environment.

United States

A 2005 report commissioned by the City of Key West Naval Properties Local Redevelopment Authority cited deep-draft vessels scouring the bottom, resuspending sediments, significantly increasing turbidity, and interfering with historic diving and fishing activities. Each year, 390 cruise ships docked in Key West.

The Key West Committee for Safer, Cleaner Ships formed in 2020, intent on balancing environmental and economic goals. The Committee's conclusions are telling.

- Recreational and commercial fishing in the Keys supports 8,000 jobs. Declining marine life populations threatens the economy far more than the loss of the cruise ship business.
- Fifty percent of visitors to the Keys are cruise ship passengers, but they only spend $32 per passenger, contributing 8 percent to the economy.
- Smaller, more environmentally friendly ships tend to cater to high-spending customers, spending an average of $550 per passenger, bringing more business to the economy without the destruction brought by large cruise ships.

On election day, Key West residents approved three referenda that ban many cruise ships from visiting.

- Cruise ships with more than 1,300 passengers are banned from docking.
- No more than 1,500 cruise ship visitors can disembark per day.
- Priority docking is given to cruise lines with the best environmental and health records.

The goal of the referenda is to reduce COVID-19 exposure, allow marine life to recover, reduce crowds, and attract higher-spending visitors.

Forty of the fifty cruise ships that docked in Key West last year will be excluded including Norwegian Cruise Line, Royal Caribbean, Disney Cruise Line, and Carnival Cruise Line. Luxury cruise lines like Regent Seven Seas, Seabourn, and Silversea will be less affected. Budget-minded cruisers will suffer.

Adapting the Cruise Ship Business Model

Cruise ships are not inherently bad, and the super-sized ship model is economically understandable. But the industry may have erred by failing to take into consideration the priorities and needs of their ports of call. And the ports of call may have been less diligent than needed to protect the health of their communities.

The path forward for the cruise industry and their ports of call is a conscious deepening of their collaboration. Complex issues need to be addressed to re-establish the value proposition. Among them, how can the ships enter ports without damaging the environment or interfering with citizens' livelihoods? And what is the carrying capacity of the port of call communities?

Mutually beneficial cruise ship and port of call relationships are possible, but achieving them will require some difficult decisions about ship size, access to onshore activities, and balancing cruise ship needs with the well-being and prosperity of port communities. The ports of call also have a responsibility to understand what is best for their communities.

THE VALUE OF TREES

Being thus prepared for us in all ways, and made beautiful, and good for food, and for building, and for instruments of our hands, this race of plants, deserving boundless affection and admiration from us, becomes, in proportion to their obtaining it, a nearly perfect test of our being in right temper of mind and way of life; so that no one can be far wrong in either who loves trees enough, and everyone is assuredly wrong in both who does not love them, if his life has brought them in his way.

John Ruskin, writer, philosopher, art critic and polymath of the Victorian era

Ecosystem Services

The Earth's forests are worth trillions of dollars. Forest value is derived not only from the timber, paper, food, drugs, firewood, and nearly 5,000 other products we consume, but from the regulating, cultural, supporting, and provisioning services they provide. Services that produce the oxygen you breathe, purify the water you drink, regulate your climate, buffer global warming, prevent soil erosion, protect you from storms, and support genetic and species diversity.

The value of trees is also realized at the local level. The trees of your neighborhood parks and that line your streets, even the trees in your backyards, are valuable assets. They ease your mind, enhance your health, reduce air pollution, increase your property value, and reduce your utility costs.

Symbolism

Trees often symbolize human or religious qualities. The Jacobins of the French Revolution planted trees in Paris and decorated them with symbols of liberty, unity, and equality. Similarly, U.S. Americans planted poplars during the Revolutionary War as symbols of growing freedom. The Sacred Tree, one of the most widely known symbols on Earth, is alternately an image of the cosmos, a dwelling place for gods and spirits, and a medium of prophecy and knowledge.

Pleasing Landscapes

The 17th Century British poet Thomas Vaughan wrote, *Sometimes thou may'st walk in groves, which being full of majestie will much advance the soul.* Science has since verified this sentiment. Trees are pleasing elements of the landscape that convey a sense of serenity and beauty. Drivers prefer parkways to less scenic highways, even if it adds time to their trip. Public garden visitors consider trees the most enjoyable aspect of their experience. And at home, people prefer views of trees to views of large open spaces.

Well-being and Health

Natural settings, of which trees are an important element, convey a sense of well-being that promotes rest and relaxation. Trees facilitate recovery from mental fatigue or psychophysiological stress. Natural and landscaped urban settings make people feel better – slower heartbeat, lower blood pressure, and more relaxed brain wave patterns - than non-landscaped urban settings.

So, too, natural settings enhance our mental and physical health. Self-esteem, happiness, educational status, and emotional development increase in at-risk children and others exposed to natural settings. Hospital patients with windows facing trees have shorter hospital stays and require less pain-killing drugs than patients with windows facing a wall.

Noise Reduction

Trees help reduce the clatter of everyday life by absorbing and blocking noise. The rustle of leaves and the wind blowing through branches creates a pleasing background sound. Planners have learned that dense, tall, and wide tree belts significantly reduce highway noise.

Cooling

Forested areas are cooler in the daytime and warmer in the nighttime than treeless areas. Trees are especially beneficial in countering the heat island effect of cities. Concrete, steel, and asphalt store thermal energy, increasing the temperature in urban areas by 3 to 10 °C compared to the countryside. Trees absorb heat as they transpire, provide shade that reduces solar radiation and reflection, channel breezes, and increase fog precipitation. The net effect is that cities with trees are 1 to 4 °C cooler than cities without trees.

Healthy Air

We breathe easier because of trees. One acre of trees produces enough oxygen for eighteen people. Conversely, trees help settle, trap, and hold particulate pollutants like ash, dust, pollen, and smoke that damage our lungs. For example, tree lined streets in Frankfurt, Germany have 25 percent less particles than streets without trees. Trees reduce other air pollutants as well. Each year, one acre of trees absorbs carbon dioxide equivalent to that produced by a car driven 21,000 to 26,000 miles. Each day, one 80 ft tree removes enough carbon dioxide to offset two single-family residences. Trees further reduce air pollution indirectly through shading. Cars in tree-shaded parking lots release less reactive organic gas emissions than cars in non-shaded lots.

Energy Savings

Strategically planted trees can result in significant energy savings. Cooling costs can be reduced from 20 to 50 percent by locating deciduous trees to shade the windows, walls, and roof of a building. One large, mature tree has the cooling effect of ten room-capacity air conditioners running 20 hours per day. Conversely, coniferous trees forming a windbreak can reduce heating costs by 10 to 50 percent. Overall, trees can produce energy savings of 20 to 75 percent over the course of a year in temperate regions. Simulations suggest that even a single, 25 ft tall tree can reduce annual heating and cooling costs for a typical residence by 8 to 12 percent.

Stormwater and Flood Mitigation

Stormwater and flood management benefits from trees. Tree canopies intercept rainfall, and organic matter on the soil surface from leaf litter and roots increases soil permeability. The net effect is that forested areas have reduced surface runoff, erosion, and sedimentation and pollution of waterways, and increased groundwater recharge. For example, in Salt Lake City the urban forest reduces stormwater runoff by 17 percent - more than eleven million gallons. In Atlanta, trees provide stormwater and flood management benefits comparable to a $2 billion reservoir.

Property Values

Closest to home, trees increase property values and enhance community economic stability. Homebuyers prefer developments with parks, gardens, and wilderness areas. Property values of landscaped homes are 5 to 20 percent higher than the value of non-landscaped homes. Office and industrial space in a wooded setting have a higher demand than sites without trees. Trees also attract businesses and tourists, and induce customers to pay up to 10 percent more for goods and services.

If trees were a publicly traded company, we would conclude they offer an outstanding return on investment. For a small initial investment and minimal operation and maintenance costs, trees enhance our health, temper our climate, clean our air, reduce our energy costs, and manage our stormwater. Invest wisely - plant a tree.

For in the true nature of things, if we rightly consider, every green tree is far more glorious than if it were made of gold and silver.

Martin Luther, priest, theologian, author,
hymn writer, and professor

IT TAKES MANY VILLAGES TO REVERSE DESERTIFICATION

The Great Green Wall promises to be a real game-changer, providing a brighter future for rural youth in Africa and a chance to revitalise whole communities. It can unite young people around a common, epic ambition: to 'Grow a 21st Century World Wonder', across borders and across Africa.

Monique Barbut, Executive Secretary of the United Nations Convention to Combat Desertification

Desertification - land degradation in arid, semi-arid, and dry subhumid areas - encompasses one quarter of the total land area of the world and affects one sixth of the population. Land use change, land use intensification, and climate change contribute to desertification.

The results of desertification are many, including reduced soil fertility, persistent droughts, lack of food and water, conflicts over dwindling natural resources, poverty, and mass out-migration. Vegetation loss exacerbates climate change and reduces biodiversity.

In 1992, the United Nations crafted an Agenda for combating desertification and drought. Programs included:

- Strengthening knowledge and monitoring systems prone to desertification.
- Intensifying soil conservation, afforestation, and reforestation activities.
- Eradicating poverty and promoting alternative livelihoods.
- Developing comprehensive anti-desertification programs within national development plans.
- Developing drought preparedness and drought relief schemes.
- Encouraging and promoting environmental education.

The United Nations declared 2010 to 2020 the "Decade for Deserts and the Fight Against Desertification".

The *Great Green Wall* initiative of the Sahel Region of Africa is an exemplar of how to reverse desertification. In the early 1970's, climate change-related droughts, population growth, and unsustainable land management practices in the Sahel led to the loss of livestock and cereal crops. Food and water were scarce, conflicts over dwindling natural resources increased, and many people migrated to Europe.

In the 1980s, the idea of a Great Green Wall surfaced. In 2007, under the leadership of the African Union, eleven countries agreed to participate. Today, an ever-expanding membership includes thirty countries. By planting trees and other plants in an 8,000 km swath across Africa, the Wall's 2030 goals are:

- Restore one hundred million hectares of currently degraded land.
- Sequester 275 million tons of carbon.
- Create ten million jobs in rural areas.

Twenty-eight million hectares of land were restored in the first 10 years, providing food security, jobs, and motivation to remain in place for the millions who live in the area. The initiative is testament to what can be accomplished when individuals and countries own a solution and work together.

NATURE OUTPERFORMS TRADITIONAL HAZARD DEFENSES

Cheap, effective, and scalable, Nature-based Solutions can help governments address the growing challenges of climate change, biodiversity loss, increased frequency of extreme weather and natural hazards as well as other human-made environmental disasters.

United Nations Office for Disaster Risk Reduction

The National Oceanic and Atmospheric Administration reports that the United States had 273 weather and climate disasters with damage costs of $1 billion or more since 1980. The total cost of these 273 events exceeded US$1.8 trillion. The number of events, costs, and deaths is increasing.

Nature is equal to or better than conventional structural approaches for hazard mitigation, and are more cost-effective. These are the conclusions of a 2020 report by the National Wildlife Federation and Allied World Assurance Company. Natural and nature-based approaches are effective against inland flooding, coastal hazards, extreme heat and drought, and wildfires.

Natural approaches to hazard mitigation leverage intact or restored ecosystems. Nature-based approaches mimic natural systems, but are designed and constructed by people. Examples include wetlands, coral reefs, dunes, floodplains, and forests. Conventional approaches to hazard mitigation are *hard* or *gray* infrastructure. Examples include embankments, dams, levees, or shade structures.

Coral reefs reduce wave energy by an average of 97% and marshlands reduce wave height by 70% or more. Along the Gulf Coast, the estimated benefit-to-cost ratio of wetland restoration for flood risk reduction is 8:1. The benefit-to-cost ratio for levees is 0.99:1. Trees reduce urban heat stress by as much as 13 percent while purifying the air.

The National Wildlife Federation and Allied World report describes natural and nature-based approaches for inland flooding, coastal hazards, extreme heat and drought, and wildfires. Recommendations for reducing damage, costs, and deaths associated with hazard events include:

- Identify, protect, and restore natural defenses.
- Mainstream the use of natural and nature-based approaches.
- Avoid development in risky environments.
- Socialize the use of natural and nature-based approaches among communities, agencies, decision-makers, and developers.
- Expand research on performance and effectiveness.
- Increase funding for natural and nature-based hazard protection.

MANGROVE FORESTS AS NATIONAL COASTAL INFRASTRUCTURE

Mangroves are rare, spectacular, and prolific ecosystems on the boundary between land and sea. They ensure food security for local communities. They provide biomass, forest products, and sustain fisheries. They contribute to the protection of coastlines. They help mitigate the effects of climate change and extreme weather events

Irina Bokova, Director-General of the UN Educational, Scientific and Cultural Organization

A recent study in Scientific Reports estimated global mangrove forests reduce property damage by $US65 billion and protect more than fifteen million people each year. Twenty-nine percent more land would be lost, 28 percent more people left unprotected, and 9 percent more property damaged if existing mangroves were eliminated. Mangroves are destroyed at a rate of 1 percent per year, higher than the rate for terrestrial forests.

About 90 percent of mangrove flood protection benefits stem from protection from hurricanes, cyclones, and typhoons. The value of mangrove reductions in regular flooding is valued at $US5 billion with one million people protected every year. Miami, United States and Cancun, Mexico avoided more than $US500 million in property damages each year because of mangroves. The flood protection benefits of mangrove forests will increase as climate change produces more intense storms and more frequent flooding events.

Avoided deaths and destruction of land and property argue for the use of hazard mitigation and disaster recovery funds for restoring and expanding mangrove forests. Large-scale restoration and expansion are possible.

Senegal's former Minister of Ecology rallied citizens to plant 152 million mangrove buds by hand. Wetland International's Indonesian Director offered people devastated by a tsunami credit to buy equipment and animals to re-establish their village if they would plant mangroves. If the trees survived for at least two years, the debt was forgiven. The restoration was successful in both instances.

In addition to flood protection, mangrove forests mitigate climate change by storing relatively high amounts of carbon and methane, stabilizing shorelines, and providing habitat for marine species and wildlife.

The World Wildlife Fund, Conservation International, the International Union for the Conservation of Nature, and The Nature Conservancy formed the Global Mangrove Alliance with the goal of increasing mangroves by 20 percent by 2030.

SAVING SEA TURTLES WITH INNOVATION AND TECHNOLOGY

For most of the wild things on earth the future must depend upon the conscience of mankind.

Archie Carr, sea turtle scientist and author

Six of the world's seven sea turtle species are classified as threatened and endangered. Threats to their survival include entanglement in fishing gear, the illegal trade of eggs and meat, coastal development, ingesting and becoming entangled in plastic and other marine debris, sea-level rise, rising temperatures, and the turtle shell trade. Nonprofit organizations, governments, and communities are working diligently to ensure that sea turtles persist.

Reducing Sea Turtle Egg Consumption

Poachers destroy more than 90 percent of sea turtle nests on Central American beaches. The eggs are collected and smuggled to cities where they are eaten as a delicacy in restaurants and bars or sold door-to-door. Some consumers believe turtle eggs improve sexual performance.

Sea turtle egg poaching is traditionally countered by community rangers willing to spend days and nights collecting scientific data and patrolling for poachers. A shortage of rangers, too much beach, too many nests, the strong demand for sea turtle eggs, the poverty of citizens in developing economies, and poor knowledge of how the eggs reach their customers limit this valiant effort.

Tracking Eggs From Poacher to Consumers

How can we overcome limited resources for protecting sea turtle nests? Kim William-Guillén of the nonprofit organization Paso Pacífico had one answer – decoy eggs equipped with GPS devices.

Kim and her colleagues identified a pliable plastic that mimicked sea turtle eggs. Using a 3D printer to fabricate decoys, they embedded the decoy eggs with GPS tracking devices. They then slipped the decoys into more than one hundred green and olive ridley turtle nests on four beaches in Costa Rica.

Poachers collected twenty-five decoy eggs and five were tracked over distances ranging from 28 m to 137 km. The decoy egg traveling the greatest distance was traced to an alley behind a Costa Rica Central Valley supermarket, and then to a private residence. You can only imagine the individual's surprise when they prepare to eat the egg.

Emblematic of Conservation Solutions

The sea turtle decoy eggs are emblematic of the creativity and innovation needed to address complex conservation challenges. Identifying trafficking patterns supports law enforcement and focuses community engagement and education efforts. The solution is potentially applicable to other egg-laying reptiles and birds and reflects an increased effort to track illegal wildlife with remote sensors.

NEW TOOLS FOR SUSTAINING WILDLIFE

Science and technology are the keys to both our longevity and our demise. Our entire existence on this planet is a double-edged sword.

Rhys Darby, actor, and comedian

Science and technology can facilitate solutions to ecosystem resilience and sustainability. Especially if driven by vision and strategy, accompanied by education, policy, and communication, implemented by critical stakeholders, and buoyed by hope.

Here are four examples of tools making a difference.

FLYING LABS ENABLE LOCAL ACTION

Established in 2015, Switzerland-based WeRobotics empowers local community humanitarian, health, development, and environmental action. What distinguishes WeRobotics is their integration of robotics, data, and artificial intelligence (i.e., intelligence demonstrated by machines), and their co-creation of the Flying Labs Network.

Flying Labs serves as local knowledge hubs in Africa, Asia, and Latin America, regions in dire need of humanitarian, health, development, and environmental solutions. Network members receive free software, discounted aerial and underwater drones, drone add-ons, access to projects and funding opportunities, and drone, data, and artificial intelligence training. The Network enables access to more than one hundred drone and data experts.

WeRobotics Flying Labs Environmental Projects

Cameroon

In Cameroon, Africa network member Map & Rank uses drones, data tools, and a family data visualization game to profile climate risk for floods, droughts, landslides, and water epidemics. Map & Rank provides capacity-building training, is profiling and modeling climate risks, and scaling alert messages to inform the public of pending disasters.

New Mexico

In New Mexico, United States WeRobotics is working with New Mexico State University to support organic pecan growers. Aphids weaken pecan trees by feeding on moisture and nutrients. The aphid's waste (honeydew) attracts sooty mold. Drones enable biocontrol by delivering aphid predator lady beetles to the trees in a low-cost and effective way.

Senegal

In Senegal, Africa drones and artificial intelligence are helping to census birds in support of the management of the Palmarin Community Nature Reserve. Training includes drone operation for collecting data and data processing using Picterra and Pix4D software. The drones operate in inaccessible areas, are more accurate than ground-based surveys, and improve decision-support management tools.

Hindu Kush

In the Hindu Kush Himalayas of western Nepal, drones repeatedly map the Ponkar Glacier, a focus of the Kathmandu University Himalayan Cryosphere, Climate, and Disaster Research Center. Himalayan glaciers are a source of water for downstream communities and serve as climate change indicators.

BEATING WILDLIFE TRAFFICKERS AT THEIR OWN GAME

Wildlife trafficking and poaching is a US$20 billion per year business for organized crime syndicates. The business model is sophisticated, encompassing bribes, acquiring location metadata from digital, online photographs, and hacking legitimate GPS tracking data from researchers and conservation managers. In recent years, trafficking has moved from physical marketplaces to an unregulated online marketplace populated with anonymous vendors and buyers.

In a 2018 study, the International Fund for Animal Welfare (IFAW) identified 5,000 advertisements on one hundred online marketplaces and social medium platforms offering nearly 12,000 endangered specimens worth US$4 million. For many species, like elephants, tigers, pangolins, rhinoceros, great white sharks, and wolves, the risk of extinction is increasing.

Fortunately, tools are being developed and deployed to defeat this sophisticated trafficking and poaching.

Hack the Poacher

Poachers rely on mobile phones and radios for navigation and to communicate with each other. Hack the Planet establishes a system of network sensors in areas targeted by poachers. The sensors detect the poachers' mobile phone signals and radio frequencies, and alert rangers to their presence.

The Hack the Poacher system monitors up to 115 square miles with only 30-40 sensors. Hack the Planet is updating the system to work with camera traps, machine learning, and satellite imagery.

Shuttering Online Markets

IFAW is working with e-commerce, technology, and social media companies to eliminate online, illegal wildlife marketplace sites. The Coalition to End Wildlife Trafficking Online collaborates with thirty-six companies throughout the world to reduce wildlife trafficking online. Collaborators include Google, Microsoft, eBay, and Tencent.

Collaboration and Innovation

The United States Department of State co-hosted Zoohackathon 2020, bringing together students, coders, developers, designers, project managers, and subject matter experts to develop innovative solutions to wildlife trafficking issues. Occurring over two and a half days, Zoohackathon consisted of five virtual, regional events across the globe. The event also examined links between wildlife trafficking and zoonotic disease i.e., diseases that pass from animals to humans.

COUNTING FISH WITH ENVIRONMENTAL DNA

An accurate assessment of fish stocks is essential for sustainable management. Trawl surveys are the common method, but are constrained by expense and replicability. Also, trawling can damage habitats and bottom organisms, potentially undermining the very fish stocks they are meant to sustain.

Recent environmental DNA (eDNA) studies illustrate the potential for dispensing with trawls, reducing costs, maintaining accuracy, and avoiding habitat and organism damage.

Baltic Sea

A 2019 study in the Baltic Sea found no correlation between eDNA concentrations and trawl biomass. Although, the study did find an association between measured concentrations of eDNA and the distributions and abundances of fish species. The authors concluded that eDNA analysis may not be directly comparable to trawl findings, but it could prove a useful supplement for monitoring fish stocks.

Atlantic Cod

A study of Atlantic Cod in the Faroe Islands region found stronger evidence of the potential for fish stock eDNA assessments. Trawl catches and eDNA detection was in accordance 80 percent of the time. Trawl biomass and catch per unit effort were significantly correlated with eDNA quantities and concentrations, respectively. The findings extend the potential application of eDNA to biomass assessments of commercially important species.

New Jersey

A study of marine fish diversity, seasonality, and relative abundance in coastal New Jersey further advance the potential of eDNA. Trawl and eDNA peak season abundance agreed for 70 percent of fish species. And eDNA agreed with monthly trawl estimates of marine fish species richness, composition, seasonality, and relative abundance.

THE INCREDIBLE SHRINKING GPS TRACKER

The ability to track animal movement is a prerequisite to conservation. Ultra-high frequency and very high-frequency radio tags are inexpensive, but the receivers must be manually tuned to frequencies and coordinates, limiting the number of locations that can be obtained. GPS tracking collars were historically limited to large, terrestrial species capable of carrying heavy batteries.

The next generation of GPS trackers are discrete, solar-powered, lightweight tags that do not impede animals, can withstand cold, heat, moisture, and dust, and use the International Space Station rather than more distant satellites. Two examples illustrate their utility.

Spoonbill Sandpiper

In 2015, there were less than 250 breeding pairs of spoon-billed sandpipers, a small shorebird that breeds in the Russian Arctic and winters on southeast Asia mudflats. The lack of information about breeding sites, migration, and wintering habitat precluded meaningful conservation efforts.

A lightweight (2 grams, about the weight of a business card) GPS tag developed by Microwave Telemetry enabled a collaboration that attached transmitter tags to three sandpipers in Jiangsu Province, China. The tags transmitted valuable location information for several months, enabling the beginning of a conservation strategy.

European Storm Petrel

The European storm petrel is a sparrow-sized seabird of the Atlantic Ocean that nests on remote islands along the coast of northwest

Europe. The breeding behavior of the petrels is reasonably known on land, but little is known about their feeding behavior at sea.

A one gram (about the weight of a standard paper clip) GPS tag enabled the acquisition of petrel location information, revealing that the petrels travel more than one hundred miles to feed for one to three days before returning to their chicks. The findings inform the designation of Important Bird and Biodiversity Areas.

THE AMUR LEOPARD RETURNS TO CHINA

The project has improved the living environment of the North China leopards and has reduced the impact of human activities on them.

Yang Xiaodong, chair of an animal protection association in Heshun county, Shanxi, China

In 2002, the Amur leopard, whose range once encompassed southeastern Russia, the Korean peninsula, and northeastern China, consisted only of a small relict population of 25 to 40 individuals in far east Russia. But beginning in November 2012, this critically endangered leopard has been repeatedly spotted in northeastern China. Since 2018, evidence of leopard breeding in Heshun County was supported by sightings of fifteen leopard cubs and an infrared camera image of a female with two cubs. Forty-four leopards were identified this past year.

The Amur leopard is a China national first-class protected species. The leopard historically occupied temperate, broadleaf forests of the Taihang Mountains in Beijing, Hebei, Shanxi, and Henan provinces. Requiring large territories to avoid food competition, Amur leopards hunt deer, badges, hares, and other animals at night. The population was decimated by deforestation and illegal hunting in the last half of the 20th century.

Afforestation is the chief reason for the Amur leopard's resurgence. Supported by national designation for protection, local governments, nonprofit organizations, and businesses have acted to sustain the population.

- The Heshun County government in Shanxi province compensates farmers US$150 to $300 for each livestock animal killed by a leopard, reducing retaliation.
- Governments modified highway routes and installed animal-safe passages.
- Chinese financial technology firm Ant Group and the Beijing-based Society of Entrepreneurs and Ecology Foundation established a leopard preserve, and promote monitoring, anti-poaching, and compensation measures.

The Amur tiger, deer, and other wildlife are benefitting from leopard conservation efforts.

DNA analysis reveals that the Amur leopard has the lowest genetic diversity of any subspecies, comparable to genetically depleted Florida panther and Asiatic lion populations. The leopard is at risk of congenital and reproductive abnormalities associated with inbreeding depression. In the wild, leopards only live for 10 to 15 years. The China population requires augmentation of captive bred individuals to increase genetic diversity and resistance to abnormalities and disease, an action supported by the international conservation community.

HOW MUCH IS ENOUGH?

We are using the equivalent of 1.6 Earths to maintain our current way of life ... One million of the world's estimated 8 million species of plants and animals are threatened with extinction ... Close to 90% of the world's marine fish stocks are fully exploited, overexploited, or depleted.

United Nations Environmental Program

The debate continues as to how much of the planet's surface must be protected to conserve biodiversity. A recent summary by Woodley et al. finds support for biodiversity conservation targets ranging from 30 to 70 percent of the land and sea. Some researchers are wary of the social acceptability of a large number. Of more concern should be getting it right.

Scientists use four approaches to determine the area required to conserve biodiversity.

1. Species area curves describe the relationship between ecosystem area and number of species. The Half Earth organization concludes that conserving 50 percent of the earth would conserve 85 percent of species. The remaining 15 percent would persist on the other 50 percent or be lost.
2. Systematic conservation planning sets targets and then selects sets of valued ecosystem components such as species or

ecosystem types. Estimates of required areas range from 25 to 70 percent for terrestrial systems, and 20 to 40 percent for marine systems.

3. Establishing the minimum area required to maintain an intact, functioning ecosystem. Analyses of the Amazon basin estimate conservation areas ranging from 60 to 80 percent of the total.
4. The Global Deal for Nature started with the goals of limiting climate change to 1.5°C and conserving biodiversity. Their 2030 target is for 30 percent of the earth to be formally protected, and an additional 20 percent designated as climate stabilization areas.

Leading with a percent of Earth to be conserved is a good start and attractive in its simplicity. The idea appeals to conservationists and relevant funding sources. But conceptual simplicity may undermine the successful execution of a percent-based plan. From a scientific perspective, conservation goals must be more specifically articulated in a way understood by diverse stakeholders. Why those goals and not others?

Then, considerable analysis is needed to determine which combination of landscape attributes and areas best achieve the goals. To some extent, we are hampered by our incomplete knowledge of cause and effect. We understand the consequences of change at the ecosystem level better than at the species level, but in each instance uncertainty is high. To our advantage, factoring in our improved understanding of ecosystem services should help and appeal to multiple stakeholders if communicated effectively.

Politics, sovereignty, and fairness pose considerable challenges to consensus and implementation. Like discussions around climate change action and biodiversity conservation, does the burden fall on less developed countries? What sacrifices should more developed countries make? Will the global community be content with sharing technologies that lessen development impacts? Or should developed countries be expected to restore critical areas already impacted?

Socioecological systems are essential to identifying, designing, implementing, and managing conserved areas. As currently constructed, regulation, protection, and education are inadequate to reverse environmental destruction and degradation. Stakeholder identification, analysis, and engagement are necessary to ensure we overcome the burden of knowledge, activate collective intelligence, and convey ownership. Too often our top-down approach ignores and disrespects the very communities needed to sustain biodiversity.

Conserving biodiversity requires a global, multi-scale, cross-disciplinary, diverse stakeholder-derived plan. A plan that draws on scientists and communities to plug gaps in knowledge. A plan based on science but informed by policy, business, and community. A plan that is effectively communicated to all stakeholders throughout the process of its development. A plan that has the support of most of the populace.

Halting environmental destruction and restoring healthy lands and waters is urgent. Good work is happening, but progress is not keeping pace with the challenge. It seems that timely success will require a single global plan created by smart, collaborative, cooperative, and sensitive people sitting in the same (virtual) room.

WELL-BEING AND PROSPERITY DEMAND THAT EVERY DAY IS EARTH DAY

The Earth is what we all have in common.

Wendell Berry, novelist, poet, essayist, environmental activist, cultural critic, and farmer

Happy Earth Day everyone!

Why one day? Every day should be Earth Day. Our well-being and prosperity are inextricably connected to the health of the Earth.

Our Health Depends on a Healthy Earth

We should begin every day committed to ensuring healthy lands and waters. If for no other reason than it is in our own best interests. Every day should end with a review of what we did or did not do to ensure a healthy Earth.

Act Out of Compassion or Self-interest

As co-inhabitants of the same ecological system, each of us bears some responsibility for ensuring everyone has food, water, and a non-toxic environment. The death of millions of people each year unable to meet basic needs should not be ignored or tolerated. Begging, illegal activities, migration, and conflict should not be everyday activities for the survival of millions, while food is wasted, and profits are reaped, by degrading resources common to all.

Human Activity has Exceeded the Earth's Nine Planetary Boundaries

We are a planet in a decline that threatens the continuance of humanity for generations to come. Success is not only short-term economic gain or the acquisition of possessions, but long-term human health and happiness, and the health of the Earth's natural resources. The Earth urgently needs a paradigm-shifting, common vision and strategy that includes explicit, enforceable roles and responsibilities for governments, nongovernment organizations, corporations, communities, and individuals.

Every day must be Earth Day, or the Earth and human well-being and prosperity will be lost.

PRINCIPLE SOURCES

We Need Healthy Lands and Waters

Intergovernmental Science-Policy Platform on Biodiversity and Ecosystem Services. 2018. Nature's Contributions to People (NCP) – Article by IPBES Experts in Science <https://www.ipbes.net/news> (Intergovernmental Platform on Biodiversity and Ecosystem Services (IPBES), Bonn, Germany.

Millennium Ecosystem Assessment. 2005. Ecosystems and Human Well-being: Synthesis. Island Press, Washington, DC.

Intergovernmental Science-Policy Platform on Biodiversity and Ecosystem Services. 2019. Global Assessment Report on Biodiversity and Ecosystem Services.

World Health Organization. 2016. An estimated 12.6 million deaths each year are attributable to unhealthy environments. https://www.who.int/newsroom/detail/15-03-2016-an-estimated-12-6-million-deaths-each-year-are-attributable-to-unhealthy-environments.

World Health Organization. 2016. An estimated 12.6 million deaths each year are attributable to unhealthy environments. https://www.who.int/newsroom/detail/15-03-2016-an-estimated-12-6-million-deaths-each-year-are-attributable-to-unhealthy-environments.

What Does the Public Think About Climate Change?

https://www.pewresearch.org/science/2020/09/29/concern-over-climate-and-the-environment-predominates-among-these-publics/

https://www.washingtonpost.com/climate-environment/americans-increasingly-see-climate-change-as-a-crisis-poll-shows/2019/09/12/74234db0-cd2a-11e9-87fa-8501a456c003_story.html

Leiserowitz, A., Maibach, E., Rosenthal, S., Kotcher, J., Bergquist, P., Ballew, M., Goldberg, M., & Gustafson, A. (2019). *Climate change in the American mind: April 2019.* Yale University and George Mason University. New Haven, CT: Yale Program on Climate Change Communication. doi:10.17605/OSF.IO/CJ2NS

Why Should the Public Be Thinking about Climate Change?

Center for Research on the Epidemiology of Disasters and the United Nations Office for Disaster Risk Reduction. 2020. Human Cost of Disasters: An Overview of the Last 20 Years 2000 – 2019.

Too Hot for Living

S. E. Perkins-Kirkpatrick and S. C. Lewis. 2020. Increasing trends in regional heatwaves. Nature Communications. https://doi.org/10.1038/s41467-020-16970-7

A Thawing Artic Requires Urgent Action

L. Field, D. Ivanova, S. Bhattacharyya, V. Mlaker, A. Sholtz, R. Decca, A. Manzara, D. Johnson, E. Christodoulou, P. Walter, K. Katuri. 2018. Increasing Arctic Sea Ice Albedo Using Localized Reversible Geoengineering. Earth's Future 6(6):882-901.

National Oceanic and Atmospheric Administration. 2018. Arctic Report Card: Update for 2018

Effects of persistent Arctic warming continue to mount. https://arctic.noaa.gov/Report-Card/Report-Card-2018.

Scripps Institution of Oceanography. 2019. Research highlight: Loss of Arctic's reflective sea ice will advance global warming by 25 years.

How Healthy are World Heritage Sites

International Union for the Conservation of Nature. 2020. Conservation Outlook Report. https://worldheritageoutlook.iucn.org/.

Save the Rainforests of the Sea

The International Coral Reef Initiative. https://icriforum.org/.

The International Coral Reef Society. http://coralreefs.org/about-icrs/welcome/.

United Nation Environmental Program: Coral Reefs. https://www.unep.org/explore-topics/oceans-seas/what-we-do/working-regional-seas/coral-reefs.

We Need Sharks and Rays

Magda Nieduzak. 2020. 5 ways in which sharks and rays help ecosystems, other species, and people. https://sharks.panda.org/news-blogs-updates/latest-news/5-ways-in-which-sharks-and-rays-help-ecosystems-other-species-and-people.

Andy Cornish. 2020. Six Solutions to Save Sharks. https://wwf.medium.com/six-solutions-to-save-sharks-19d15cfb2e28

Squid to Die For

NBC News. 2020. The deadly secret of China's invisible armada. https://www.nbcnews.com/specials/china-illegal-fishing-fleet/index.html.

A Path to Healthy Lands and Waters

Intergovernmental Science-Policy Platform on Biodiversity and Ecosystem Services. 2019. Global Assessment Report on Biodiversity and Ecosystem Services.

U.S. Department of Education. 2019. https://www2.ed.gov/about/overview/mission/mission.html.

Daniel Pink. 2006. A Whole New Mind. Riverhead Books (Penguin Group). New York.

Grow Trust in Science By Growing Scientists

Pew Research Center. 2022. American trust in scientists and other groups declines. February 15. Report. https://www.pewresearch.org/science/2022/02/15/americans-trust-in-scientists-other-groups-declines/

Sustainable Investments Are Good for Everyone

Bridget Hoffman, Tristany Armangue i Jubert, and Eric Parrado. 2020. The Business Case for ESG Investing for Pension and Sovereign Wealth Funds. Inter-American Development Bank.

An Award-winning Payment for Environmental Services Program

National Forestry Financing Fund website https://www.fonafifo.go.cr/es/servicios/pago-de-servicios-ambientales.

Restoring the South Florida Ecosystem

Comprehensive Everglades Restoration Plan. 2000. https://www.nps.gov/ever/learn/nature/cerp.htm.

Everglades Forever Act. 1994. https://floridadep.gov/eco-pro/eco-pro/content/everglades-forever-act-efa.

The Crime of Ecocide

Team Ecohustler. 2020. French citizens' assembly votes to make ecocide a crime. https://ecohustler.com/culture/french-citizens-assembly-votes-to-make-ecocide-a-crime.

Catholic News Service. 2019. Catechism will be updated to include ecological sins, pope says. https://cnstopstories.com/2019/11/15/catechism-will-be-updated-to-include-ecological-sins-pope-says/.

Stockholm Resilience Institute. Undated. The nine planetary boundaries. https://www.stockholmresilience.org/research/planetary-boundaries/the-nine-planetary-boundaries.html.

Community-based Conservation

Etienne and Beverly Wenger-Trayner. 2015. Communities of practice: A brief introduction. 15 April. https://wenger-trayner.com/introduction-to-communities-of-practice/.

Ronald Heifetz, Alexander Grashow and Marty Linsky. 2009. The Practice of Adaptive Leadership. Harvard Business Press. Cambridge, Massachusetts.

Don Kent. 2022. *Leadership Practices for Healthy Lands and Waters*. Healthy Lands Publishing.

Look to Indigenous Peoples to Sustain the Planet

United Nations Economic and Social Council. 2019. Conservation and the rights of Indigenous peoples. https://documents-dds-ny.un.org/doc/UNDOC/GEN/N19/036/91/PDF/N1903691.pdf?OpenElement

Peter Veit and Katie Reytar. 2017. By the Numbers: Indigenous and Community Land Rights. World Resources Institute. https://www.wri.org/insights/numbers-indigenous-and-community-land-rights

Expert Recommendations for Reducing Pandemic Risk

Josef Settele, Sandra Díaz and Eduardo Brondizio and Dr. Peter Daszak. 2020. COVID-19 Stimulus Measures Must Save Lives, Protect Livelihoods, and Safeguard Nature to Reduce the Risk of Future Pandemics. https://ipbes.net/covid19stimulus.

Ensuring Food Security During a Pandemic

United Nations. 2020. Policy brief: The impact of COVID-19 on Food Security and Nutrition. https://unsdg.un.org/resources/policy-brief-impact-covid-19-food-security-and-nutrition

COVID-19 Forces Islands to Rethink Food Security

Lisa Cornish. 2020. How COVID-19 and climate are encouraging islands to invest in food security.
https://www.devex.com/news/how-covid-19-and-climate-are-encouraging-islands-to-invest-in-food-security-97295.

European Union Food Security Policy

European Commission. Common Agricultural Policy. https://agriculture.ec.europa.eu/common-agricultural-policy_en.

European Commission. Farm to Fork Strategy. https://food.ec.europa.eu/horizontal-topics/farm-fork-strategy_en#:~:text=The%20Farm%20to%20Fork%20Strategy%20aims%20to%20accelerate%20our%20transition,reverse%20the%20loss%20of%20biodiversity.

Food Security Requires Sustainable Agriculture

B.P. Oberč and A. Arroyo Schnell. 2020. Approaches to Sustainable Agriculture. IUCN European Regional Office.

Why What We Eat Matters

Brent Loken. 2020. Bending the Curve: The Restorative Power of Planet-based Diets. World Wildlife Fund. https://drive.google.com/file/d/1pVpYTNQKAE_izp8tsVK3tQ4fNWPgoQIz/view.

Independent Expert Group. 2020. Global Nutrition Report: Action on Equity to End Malnutrition. file:///C:/Users/docdk.000/Downloads/2020_Global_Nutrition_Report_2hrssKo.pdf.

Jonathan Woetzel, Dickon Pinner, Hamid Samandari, Hauke Engel, Mekala Krishnan, Nicolas Denis, and Tilman Melzer. 2018. Will the world's breadbaskets become less reliable. McKinsey Global Institute. https://www.mckinsey.com/capabilities/sustainability/our-insights/will-the-worlds-breadbaskets-become-less-reliable.

McKinsey & Company. 2019. The future of food: Meatless? chrome-extension://efaidnbmnnnibpcajpcglclefindmkaj/https://www.mckinsey.com/~/media/McKinsey/Featured%20Insights/The%20Next%20Normal/The-next-normal-The-rise-of-Alternative-proteins-February-2020-Collection.

How to Regenerate Ecosystems While Growing Food

Rodale Institute. https://rodaleinstitute.org/why-organic/organic-basics/regenerative-organic-agriculture/.

Regeneration International. https://regenerationinternational.org/2017/02/24/what-is-regenerative-agriculture/.

reNature. https://www.renature.co/.

Syngenta Good Growth Plan. 2020. chrome-extension://efaidnbmnnnibpcajpcglclefindmkaj/https://www.syngenta.com/sites/syngenta/files/presentation-and-publication/updated/Annual%20general%20meeting%202015/michel-demare-agm-presentation-2015-en.pdf.

Seth J. Theuerkauf, James A. Morris Jr., Tiffany J. Waters, Lisa C. Wickliffe, Heidi K. Alleway, Robert C. Jones. A global spatial analysis reveals where marine aquaculture can benefit nature and people. PLOS ONE. https://journals.plos.org/plosone/article?id=10.1371/journal.pone.0222282.

Adapting Food Systems to Climate Change

Mbow, C., C. Rosenzweig, L.G. Barioni, T.G. Benton, M. Herrero, M. Krishnapillai, E. Liwenga, P. Pradhan, M.G. Rivera-Ferre, T. Sapkota, F.N. Tubiello, Y. Xu, 2019: Food Security. In: Climate Change and Land: an IPCC special report on climate change, desertification, land degradation, sustainable land management, food security, and greenhouse gas fluxes in terrestrial ecosystems.

Committee on World Food Security. 2014. Global Strategic Framework for Food Security and Nutrition. chrome-extension://efaidnbmnnnibpcajpcglclefindmkaj/https://www.unscn.org/files/cfs/GSF_Version_3_EN.pdf.

National Oceanographic and Atmospheric Organization. Understanding Our Changing Climate. https://www.fisheries.noaa.gov/insight/understanding-our-changing-climate.

Building Urban Resilience to Climate Change

LANDPROCESS. http://www.landprocess.co.th/.

Porous City. http://www.porouscity.org/.

Destinations Push Back on Cruise Ships

Amy Lupica. 2020. Key West Votes to Ban Large Cruise Ships Citing Economic and Environmental Impact. Our Daily Planet. https://www.ourdailyplanet.com/story/key-west-votes-to-ban-large-cruise-ships-citing-economic-and-environmental-impact/.

Talia Lakritz. 2020. 7 places being ruined by cruise ships. Insider. https://www.insider.com/cruise-ships-environmental-impact-tourism-2019-9.

It Takes a Village to Reverse Desertification

The Great Green Wall. https://www.greatgreenwall.org.

Natural Defenses Outperform Traditional Hazard Approaches

P. Glick, E. Powell, S. Schlesinger, J. Ritter, B.A. Stein, and A. Fuller. 2020. The Protective Value of Nature. A Review of the Effectiveness of Natural Infrastructure for Hazard Risk Reduction. Washington, DC: National Wildlife Federation.https://www.nwf.org/protective-value-of-nature?_ga=2.195245867.609906885.1664390482-123124046.1664390482.

Mangrove Forests as National Coastal Infrastructure

Maricé Leal and D. Mark. 2022. The State of the World's Mangroves. https://www.mangrovealliance.org/mangrove-forests/.

Saving Sea Turtles with Technology and Innovation

Helen Pheasey, David L. Roberts, Daniela Rojas-Cañizales, Carmen Mejías-Balsalobre, Richard A. Griffiths, and Kim Williams-Guillen. 2020. Using GPS-enabled decoy turtle eggs to track illegal trade. Current Biology 30, R1063–R1070. DOI:https://doi.org/10.1016/j.cub.2020.08.065

Paso Pacifico website https://pasopacifico.org/project/saving-sea-turtles/.

New Tools for Sustaining Wildlife

WeRobotics website https://werobotics.org/.

Nova. The 21st Century Threat to Wildlife is Cyberpoaching. https://www.pbs.org/wgbh/nova/article/21st-century-threat-wildlife-cyberpoaching/.

IFAW. Disrupt Wildlife Cybercrime. https://www.ifaw.org/resources/disrupt-wildlife-cybercrime.

Hack the Planet. https://www.hacktheplanet.com/.

Steen WilhelmKnudsen, Rasmus BachEbert, MartinHesselsøe, FranziskaKuntke, JakobHassingboe,Peter BondgaardMortensen, Philip FrancisThomsen, Eva EgelyngSigsgaard, Brian KlitgaardHansen, Einar EgNielsen, and Peter Rask Møller. 2018. Species-specific detection and quantification of environmental DNA from marine fishes in the Baltic Sea. Science Direct. https://www.sciencedirect.com/science/article/abs/pii/S0022098118302168.

Ian Salter, Mourits Joensen, Regin Kristiansen, Petur Steingrund, and Poul Vestergaard. 2019. Environmental DNA concentrations are correlated with regional biomass of Atlantic cod in oceanic waters. Communications Biology. https://www.nature.com/articles/s42003-019-0696-8.

Mark Y Stoeckle, Jason Adolf, Zachary Charlop-Powers, Keith J Dunton, Gregory Hinks, and Stacy M VanMorter. 2021. Trawl and eDNA assessment of marine fish diversity, seasonality, and relative abundance in coastal New Jersey, USA. ICE Journal of Marine Science. https://academic.oup.com/icesjms/article/78/1/293/6017541.

Rebecca Lee. 2016. The world's smallest satellite tag helps one of the world's rarest species! Saving the Spoon-billed Sandpiper. https://www.saving-spoon-billed-sandpiper.com/2016/10/all/the-worlds-smallest-satellite-tag-helps-one-of-the-worlds-rarest-species/.

Mark Bolton. 2020. Paperclip-sized tracker reveals the intrepid journey of UK's, smallest seabird. http://www.surfbirds.com/community-blogs/blog/2020/12/02/paperclip-sized-tracker-reveals-intrepid-journey-of-uks-smallest-seabird/.

The Amur Leopard Returns to China

China Daily/Asian News Network. 2020. Rare leopards reappearing in habitats near Beijing. September, 21. https://newsinfo.inquirer.net/1337907/rare-leopards-reappearing-in-habitats-near-beijing.

O. Uphyrkina, D. Miquelle, H. Quigley, C. Driscoll, and S. J. O'Brien. 2002. Conservation genetics of the far eastern leopard. The American Genetic Association. 93:303-311. https://academic.oup.com/jhered/article/93/5/303/2187228.

How Much is Enough

Stephen Woodley, Harvey Locke, Dan Laffoley, Kathy MacKinnon, Trevor Sandwith, and Jane Smart. 2019. A Review of evidence for area-based conservation targets for the post-2020 global biodiversity framework. Parks 25.2, November.

Well-being and Prosperity Demand That Every Day is Earth Day

Stockholm Resilience Institute. Undated. The nine planetary boundaries. https://www.stockholmresilience.org/research/planetary-boundaries/the-nine-planetary-boundaries.html.

INDEX

ACKNOWLEDGEMENTS

Any book, especially a compilation of articles, is only possible due to the work of countless individuals. Many are recognized in the section on Principal Sources.

Another source of information and inspiration are many LinkedIn connections and followers that educated me through their articles, comments, and messages. There is untold value in a social media platform replete with committed professionals, a far too rare occurrence these days.

We are all products of our familial and professional culture. I'm grateful to everyone over these last 30 years who has taught me, formally or informally, about science, policy, and the importance of integrity, competence, and enabling others.

ABOUT THE AUTHOR

Don is the principal and founder of Healthy Lands. For more than 30 years, Don has worked in socioecological systems to promote healthy, resilient, and sustainable lands and waters that support people and ecosystems.

Don has executed conservation initiatives in the United States and abroad with competencies in scientific research, strategy, planning, policy, management, communication, and leadership. He has 15 years of leadership experience in the nonprofit sector, and another 15 years in the private and government sectors.

Don is a Ph.D. in Ecology, Evolution, and Behavior, an AWAI-trained copywriter, and a member of the Professional Writers' Alliance. Career stops include an international consulting firm, a global entertainment company, a state government, and a global biodiversity nonprofit. Don is the primary author and editor of *Applied Wetlands Science and Technology* and *Leadership Practices for Healthy Lands and Waters.*

Made in the USA
Monee, IL
04 November 2026

67380067R00115